Handbook of Coil Winding

Jürgen Hagedorn • Florian Sell-Le Blanc •
Jürgen Fleischer (Eds.)

Handbook of Coil Winding

Technologies for efficient electrical wound products and their automated production

Springer Vieweg

Editors
Jürgen Hagedorn
Aumann GmbH
Espelkamp, Germany

Florian Sell-Le Blanc
Aumann GmbH
Espelkamp, Germany

Jürgen Fleischer
Karlsruher Institut für Technologie
wbk Institut für Produktionstechnik,
Karlsruhe, Germany

Additional material to this book can be downloaded from http://extras.springer.com.

ISBN 978-3-662-57195-8
DOI 10.1007/978-3-662-54402-0

ISBN 978-3-662-54402-0 (eBook)
ISBN 978-3-662-54403-7 (MyCopy)

Softcover reprint of the hardcover 1st edition 2017

Printed on acid-free paper

This Springer Vieweg imprint is published by Springer Nature
The registered company is Springer-Verlag GmbH Germany
The registered company address is: Heidelberger Platz 3, 14197 Berlin, Germany

Preface

The requirement of energy efficiency for modern electrical devices is a key challenge in product development. In addition to electronic control approach, coil winding technology offers ways of fulfilling these requirements. This is the motivation behind the efforts to create a standard reference for coil winding technology.

Coils have been around for decades and they are used in a variety of applications, for example: ignition coils in passenger cars, magnet coils for valves, electric motors or induction stoves. Cheaper semiconductors enable coils to be used in electronic devices like smartphones, sensor coils on RFID chips or proximity sensors in automation technology. In the past, coil winding was a process performed by hand or using simple machines. Today, it is an interdisciplinary field combining material sciences, process technology, mechanical engineering and control as well as electrical engineering. The emerging products show high distribution and penetration in many areas of technology in our everyday life. Winding technology products can be found anywhere, ranging from home appliances to electronics and highly efficient motors for automation solutions, or mobile applications. This large variety of applications and the different boundary conditions results in various technological trends. Primarily, the increasing requirements for energy efficiency and efficiency rates of inductive components should be stated. As a consequence, products are constantly optimised to fulfil the requirements, exerting high demands on coil winding technology. Due to the growing electrification of consumer goods in homes, private and industrial areas, supported by cheap electronics, the products' controllability is improved and losses can be reduced. This increases the significance of winding technology. With the first industrial implementation of coils, the methods, processes and machines initially used in the textile technology were applied to the production of coils. Until then, coils were often handmade in large factories. The requirements for winding technology were limited, because neither the products, like electric motors, nor the manufacturing itself were optimised. With the onset and the political acceleration of electro mobility motivated by the limitations of fossil fuels and resources, winding technology has risen in importance and has become a key expertise in electrical machine engineering. In the light of these developments, the aim of this book is to provide an introduction to coil winding technology, describe the state of the art and help overcome the current challenges in the development of winding goods and production technology. Therefore, the target group consists of engineers, who design inductive products or production technology, develop electric motors or work in the field of production technology for motors. In education, this book can help mechatronics technicians or engineering students at university to understand winding technology. Employees in the areas of procurement or logistics for inductive production technology, and operators of winding machines can also benefit from the consolidated knowledge in this book. In terms of its content, this book first covers winding goods and their properties relevant to winding technology. As a key element of winding technology, the copper wire's processing properties define process limits and the achievable product quality, and is the subject of the following chapter. The resulting automation solutions will then be presented based on the different types of windings and the available winding processes. Chapter 1 begins with an overview of the variety of inductive components, their applications and specifications, which will be described in their construction and design. The introduction of the properties relevant to the manufacturing of inductive components and a treatise of the use in electric motors with different winding patterns concludes this chapter. The structure, manufacturing and processing properties of copper wire are covered in Chapter 2. Chapter 3, which constitutes the core of this book, gives an introduction to winding processes as well as adjacent processes, such as connecting wires or assembling segments. As the final section of the book, Chapter 4 describes in detail the following processes with regard to the machine and control technology used. While referring to different levels of automation and respective production scenarios the technical implementation is illustrated by presenting common technical automation solutions. Following this, in the context of process control engineering, strategies for quality control and necessary testing technology are presented with reference to critical component properties. Practical aspects of diagnosis and prognosis of process errors, as well as particularities of winding machines in terms of operation and maintenance are covered at the end.

In conclusion, the aim of this book is to provide insights into production oriented design and production of winding goods as well as adjacent processes for manufacturing inductive electronics and electrical machine construction.

Table of contents

Introduction 1

1.1 Introduction to coil winding technology

Figure 1.1 Linear winding machine with wire guide, wire cutter and auxiliary pins

The purpose of this chapter is to give the reader insights into the superior process chain in the production of coils. Various manufacturing processes will be presented and coil variants with their physical properties will be discussed. In particular, the implications of different bobbin designs for winding technology will be discussed for different types of coils. At the end of the chapter there will be a basic introduction to electrical motors, their principles of operation as well as applications and winding systems.

The purpose of this section is to classify winding technology within manufacturing processes according to the German industry norm DIN 8580. It becomes clear that winding technology is an assembly process based on forming. The basic procedure of winding technology as part of a process sequence will be classified and key terms will be introduced.

This section provides a basic understanding of winding technology and coils.

1.1.1 Definition of coil winding technology

Coil winding technology means to wind an electrical conductor to a compact structure, a coil. Thereby, geometric structures develop by single or multiple windings, which may be layered in single or multiple layers. Coil applications can be divided into actuators, like electromagnets, electric motors and moving coils, as well as sensors, like antennas, microphones and other measuring devices. In both cases an electrical conductor, usually an enamelled copper wire, is wound onto a bobbin. The types of manufacturing range from bending of micro-technical coils with wire gauges of 12 μm to massive forming of conductor skeins with several square centimeters cross section for power plant generators. The windings in electric motors can be found in the stator and, depending on the type of motor, also in the rotor and may consist of individual coils. Winding technology has an influence both on the electromagnetic design and in manufacturing, especially on the properties of wound electromagnetic components. Power density is a priority, which, depending on the application, can be traced back to the absorption or submission of a force, a field or a magnetic flux. Other properties, such as insulation rigidity, must ensure the insulation rigidity of the component for a specific operating voltage. Since energy efficiency requirements are increasing rapidly, the development of electromechanical assemblies is becoming more and more challenging. Winding technology is divided into groups according to the type and geometry of the coils; these are presented in Sections 1.2.3 and 1.2.4. The winding process is part of a process chain and is always accompanied by upstream and downstream processes. The general process chain is described in Section 1.1.3. An introduction to common upstream and downstream processes is given in Section 3.4. Section 1.2 provides an overview of different winding products with regard to the wire gauge used.

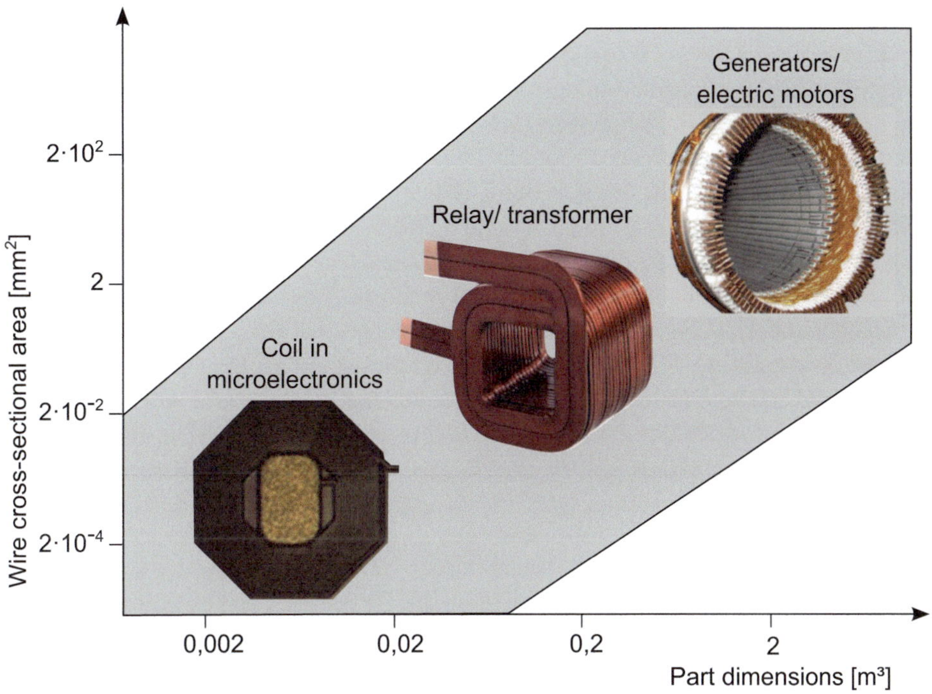

Figure 1.2 Winding technology products: microelectronic coils, relays, electric motors (Picture: F&H Porter Novelli, schwa-medico GmbH, momac Group)

Winding technology is a forming based assembly process to produce winding goods. The characteristic feature is the relative motion between the workpiece holder, which is called winding tool in coil winding technology, and the wire guidance, which is called wire guide in coil winding technology.

1.1.2 Delimitation of winding technology from other manufacturing processes

Every manufacturing process is classified by DIN 8580. According to DIN 8580, coil winding is classified as part of the main group assembly and categorised into the subgroup 4.5 Joining through shaping (DIN 8593-5) or 4.5.1 Joining through shaping of wire-shaped parts The exact group definition according to DIN can be found in 4.5.1.6 and specifies a "winding with wire to join a core with a wire by continuously bending around the core". The mentioned core is called winding carrier in winding technology and can either be a bobbin or a winding tool (Figure 1.3).

Therefore, the basic manufacturing process to create a turn is forming by bending. Forming generally describes the change of shape of a solid object while keeping its mass and material cohesion. While deforming, mechanical stresses occur in the component, which can be classified by their type of stress in the forming zone. These types of stress are grouped in five main groups: pressure forming (DIN 8583), tensile compression forming (DIN 8584), tensile forming (DIN 8585), bending forming (DIN 8586) and shear forming (DIN 8587). Aside from winding technology, bending forming is often used in the manufacturing of rods, profiles and pipes made from sheet metal. For the use of forming processes, the material involved, in this case the electrical conductor, needs to withstand the type and extent of the deformation without damage. In this context, it describes a permanent change of shape, a so-called plastic deformation. Within forming technology there is a difference between cold and hot forming depending on the material's temperature, which is typically between 500 and 800 °C. The processes of winding technology are classified as cold forming. When using cold forming the material's microstructure tends to become more rigid and the tensile strength increases, while the elongation at break and the tenacity decrease. This is called strain hardening.

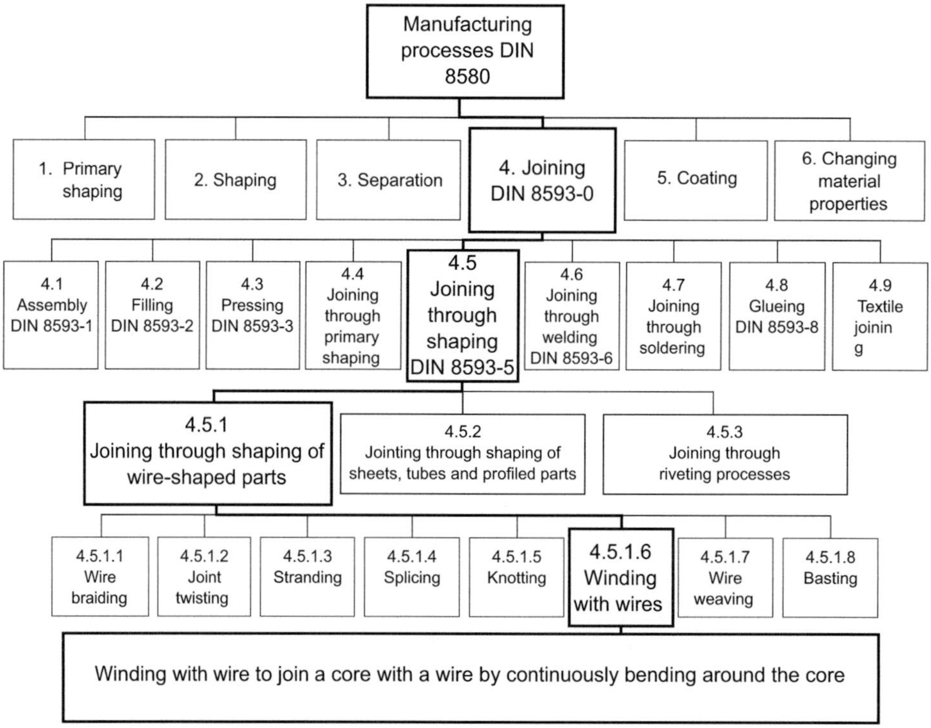

Figure 1.3 Categorization of the manufacturing process: "Winding with wire" according to German industry norm DIN 8580

Therefore, the ability to change the shape of the conductor in coil winding technology is lower compared to hot forming and is also determined by the material used. The surface of the cold formed material remains smooth and the influence of the forming speed on the change of shape is low. The mechanical properties of the commonly used winding materials will be discussed in terms of their forming properties in the Sections 2.1.1 and 2.1.2. Wire winding has positioned itself as the most commonly used coil manufacturing process. However, depending on the application and the product design, coils may also be manufactured using different processes. In this context, primary forming processes for very small geometries, as they are used in micro-electronics, or separation processes for larger geometries for use in areas with high-power currents, should be mentioned. The main characteristic of primary forming processes is the transition of a formless object to a shaped one, as for example in casting. A detailed discussion of the application background is given in the section about coil types (Section 1.2.3).

Winding is only one of many manufacturing processes for the production of winding goods. The choice of a suitable process depends on the product design. However, since the winding process is the favored process for a great number of winding products, the focus of this book is on this manufacturing process.

1.1.3 Typical process chain for the production of winding goods

The wide range of applications for winding products has led to a great range of variants and therefore different process principles in production. All winding processes for winding goods share the same general procedure of production processes, as shown in Figure 1.4.

In the case of body coils, the bobbin must first be supplied. Smaller bobbins, including those used in electronics, can be supplied as bulk goods, while bigger bobbins may be supplied by pallet. To separate the bobbins from the bulk there are various processes available, such as spiral conveyors for smaller bobbins or robotics with image processing for larger parts. Depending on the product design, upstream processes like the electrical insulation of the winding base, the area of the bobbin where the turn rests, or mechanical processing may take place. The insulation could be applied by spraying the metal bobbin with plastic, or fitting insulation materials, such as insulation paper. After mounting the bobbin on the winding tool other pre-assembly steps may take place depending on the product design. These can be used as contacting elements to contact the coil or as wrapping pins for the termination, when the wire is mechanically fixed to the bobbin. The last optional step between preparation and the actual winding process is testing for winding readiness. On the one hand, this verifies whether the contacts and insulation needed to ensure the coil's functionality have been faultlessly fitted. On the other hand, the accuracy of the bobbin's shape is checked to ensure a faultless winding process.

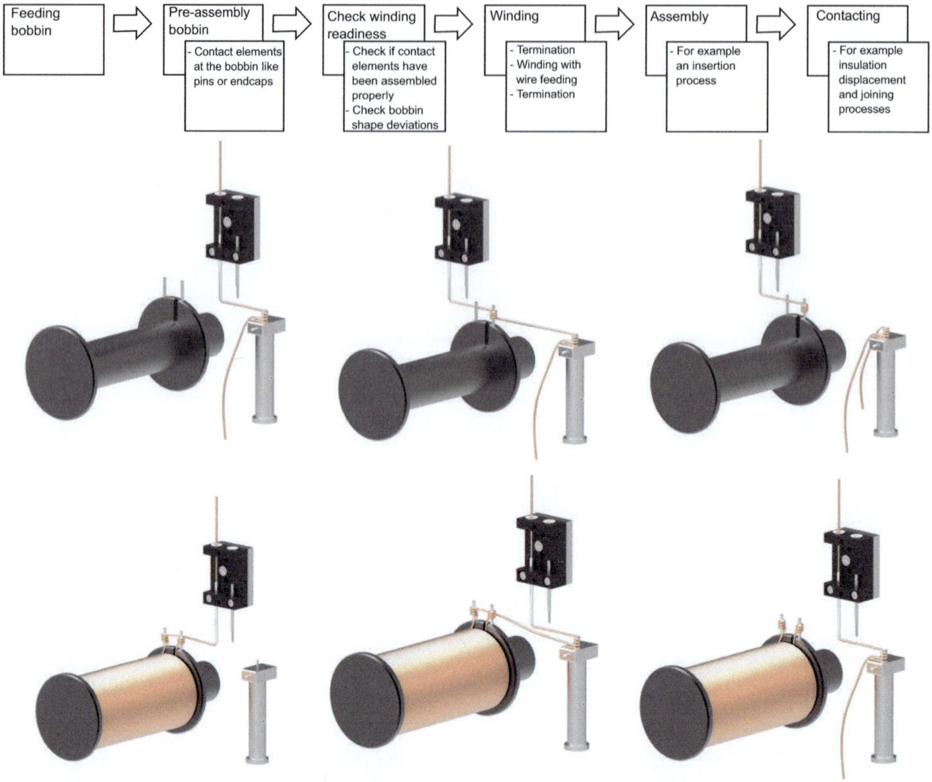

Figure 1.4 General process chain for coil manufacturing

In case of bodiless coils these steps are omitted and the winding is applied to a winding tool mounted in the winding machine. The winding process itself consists of three phases. Initially, the wire is terminated on the bobbin, which means that it is mechanically fixed on an element of the bobbin or the winding tool, before the actual winding begins. Subsequently, the other end of the wire is also terminated onto the bobbin or winding tool. An exception to this is the follow-up winding process, where the bobbin is conveyed to the next processing station without cutting the wire (Section 3.3.5). The free wire lengths, which are a result of the terminating are called start and end wire. A continuous supply of the winding good is necessary to be able to apply the winding. To ensure this, the conductor, which is provided using supply spools, is fed to a braking system through guiding elements.

Its task is to provide a defined tensile force. In addition, balancing elements are provided, which compensate the variations of the tensile force and can retrieve a loose, tension-free wire, as shown in Figure 1.5. The wire's position on the bobbin or winding tool is determined by a wire guide. As a result of the influence of the bending forming the wire fits into its intended position, as described in the previous section.

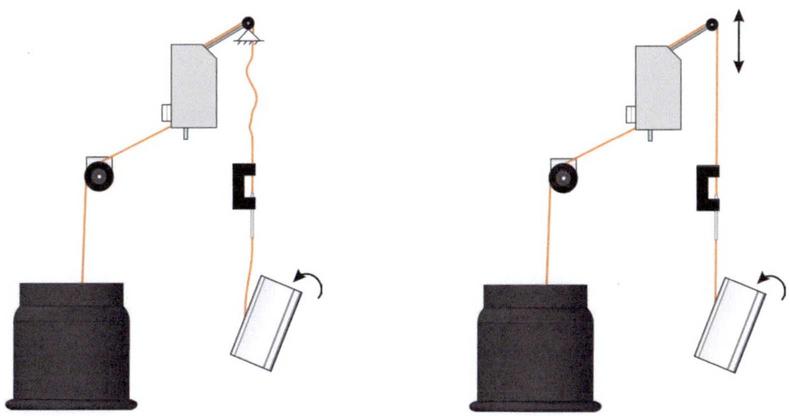

Figure 1.5 Schematics of a loose wire and its correction through a balancing arm

For the following processes the coil is taken out of the machine and is supplied to the next processing step by hand or machine. Typical downstream processes are skinning and contacting the wire to the usually already assembled contacting elements. There are a great number of process variants for this task, which need to be adapted to the individual winding task. For more complex product designs, single coils are usually joined and contacted to a whole winding design. Various processes exist here too. The most commonly used process is the insert technology for electric motor stator windings, as described in Section 3.3.8. To ensure the stability of bodiless coils the baking of the self-bonding wire is a typical downstream process. Alternatively, these air coils may be fixed or wrapped with tape. It ensures the already existing shape accuracy and enables the winding to be assembled in the intended application. In the end, the coils are stored or palletised for further assembly or transport.

The above mentioned process chain for providing winding good applies to most of the applications. The manufacturing of coils depends on the product, and individual steps may be added or left out.

1.2 Introduction to coil technology

Figure 1.6 Variety of coil types

The purpose of the first section is to differentiate different types of coils according to their manufacturing processes. In the following, the focus of this book is on coils, which are manufactured using winding technology. In addition to resistance and inductance, the other physical basics of coils as well as their properties during operation will be discussed. The following section gives an overview of existing types of coils and their specific bobbins. The background of applications, shapes and product specific characteristics will be characterised. After the shapes, the bobbin materials will be classified with regard to the different areas of requirements. Particular attention will be given to the properties of synthetic materials. A discussion of specific requirements for different types of coils follows at the end of the section.

This section, therefore, is relevant to readers, who deal with the development and design as well as the production of coils.

1.2.1 Definition of coils

Coils are geometric structures consisting of one or multiple electrically insulated conductors, which generate an electric field when a current flows, or detect a magnetic field by measuring an induced voltage. This geometric structure is called winding and can be created with different geometries and different schemes. A detailed introduction to these schemes is given in Section 3.1.2. The conductor is usually made from copper, although other materials may be used (Sections 2.1.1 and 2.1.2). This conductor is mounted to a winding carrier. For air coils this is the winding tool, while for body coils a bobbin is used (Section 1.2.4). As for the manufacturing processes, the production of the conductor differs depending on the type of coil, as described in Section 1.1.2. For the manufacturing of coils a variety of processes may be used; these are shown in Table 1.1. Besides typical coil winding technology there are other manufacturing processes, which may be suitable for production depending on the type and shape of the coil. Currently, the state of research is investigating the suitability of various primary forming coil manufacturing processes as an alternative to the winding process. For compact motor coils, which need to be manufactured very closely to the desired final shape, Fraunhofer IAM is investigating coil casting [Gro-14]. The advantage of manufacturing using primary forming is the low material stress during the production process and the ability to manufacture the parts very closely to the desired final shape. However, the winding carrier must be designed to enable collision-free assembly. Another primary forming process is already used today for coils in micro-electronics.

Coils manufactured using screen printing are most often used for sensory applications or RFID chips (radio-frequency identification), a detection mechanism based on radio-frequencies. Approaches like printing motor coils have also been tested in research [Bra-12]. Mass production of coils for applications in entertainment and communication electronics is based on a separation process from the fourth main group in DIN 8085.

Table 1.1 Manufacturing processes for coils and their application

Manufacturing process	Application
– Primary shaping: casting	Motors with single teeth
– Coating: printing [Br-12]	Micro actuators
– Coating and stripping: lithography (circuit board)	Signal processing – microelectronics
– Shaping through machining	High current applications (up to 1 kA)
– Shaping: conductor/hairpin coils	Starter/generator, fully-electric drives, high current drives
– Winding: coil winding technique	Motors, sensors, electronic, and more

Here, the coil's structure is etched from a circuit board coated with copper. The desired structure is protected by applying an acid resistant film using a lithographic process. This process chain is covered sufficiently in other literature and will therefore not be described in detail [Hil-02]. For coil applications with heavy currents the conductor diameters need to be very large. These larger mechanical structures are either assembled from smaller segments or milled from bulk material depending on the structure and application. Because of the wide prevalence of electric motors in various applications, a great variety of product designs and variants have been established. Nowadays, the complex structure of conventional distributed windings is achieved by assembling individual coils of winding wire. Alternatively, other assembling processes have been established on the basis of conductor rods, called hairpins. They are assembled as well, but manufactured using bending separation processes rather than the winding process. Typical coil shapes and their manufacturing processes are displayed in Figure 1.7. In the course of this book emphasis will be placed on coils made from winding wire.

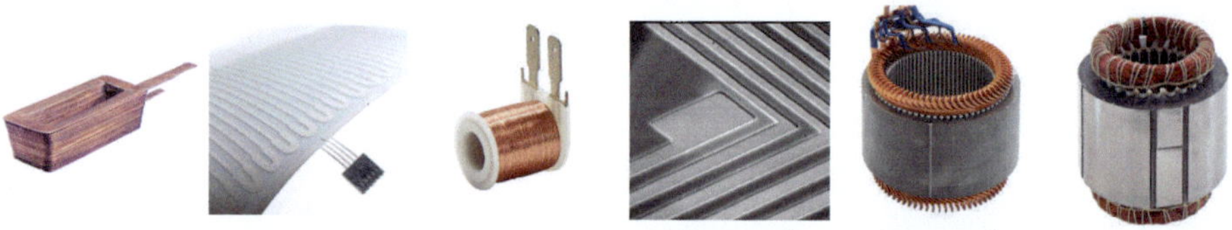

Casted coil | Printed coil | Wound coil | Micro-coil from lithography | Mounted hairpin / bar winding | Inserted winding

Figure 1.7 Samples of different coil shapes and their respective manufacturing method (Picture: Fraunhofer IFAM, Freudenberg, Department of Electrical Engineering, University of Palermo and CRES, Monreale, Italy)

In addition to the production of coil parts in various shapes, coil winding technology is also used to manufacture resistive and non-electrical parts. However, these only play a minor role. Based on the outlined variety of coil applications and designs manufacturing is faced with a wide range of challenges. These include different wire geometries and dimensions, winding schemes, bobbin shapes or power densities of coils.

1.2.2 Physical principles

The knowledge of electrical and magnetic field characteristics is essential for a basic understanding of electrical coils and electrical motors. Thus, the key physical quantities are introduced in this subsection. In general, a magnetic field can be created by the effect of a current flowing through a conductor or by certain materials. These materials have the ability to remain in place permanently once their Weiss domains have been oriented by an external magnetic field. This process is called magnetization (Figure 1.8). A variety of magnets emerge from this depending on which material is magnetised. For example, ferromagnetic materials are used also for permanent magnets. Paramagnetic materials create their own magnetic field as long as they are exposed to an outer external magnetic field. The magnetic orientation matches the outer field. Diamagnetic materials weaken an external magnetic field by creating a magnetic field with opposing poles. Field lines are used to illustrate magnetic fields. Field lines are always closed loops and never cross one another because of the separation of field sources between the North and South Poles. The following arrangement is applicable for the orientation of magnetic field lines: inside the magnetic object the field lines run from South to North Pole, while outside the object they run from North to South Pole.

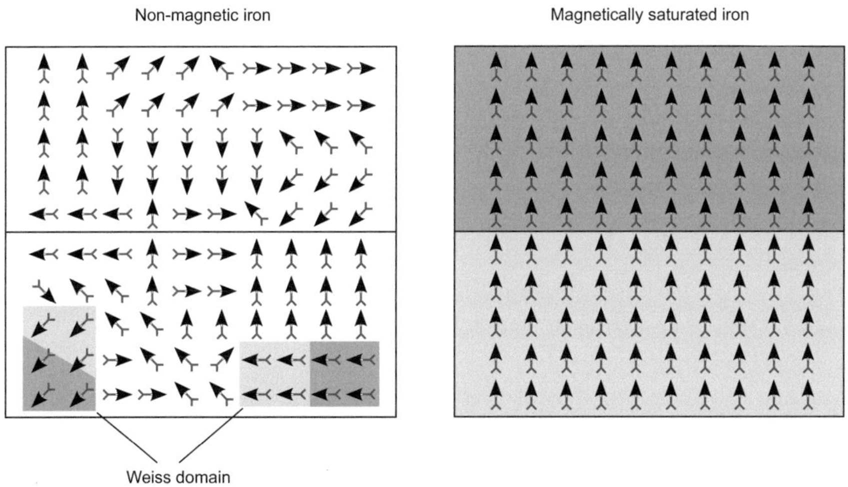

Figure 1.8 Physical effects of magnetization: Redirection of the Weiss domains

Electrical resistance

The ability to conduct an electrical current is a material depending property. The electrical resistance R, given in the unit ohm or volt/ampere, is a measure of the electrical current resulting from applied electrical voltage. The existing resistance of a round electrical conductor can be described using the following formula (1.1).

$$R = \rho \cdot \frac{l}{A} = \rho \cdot \frac{4 \cdot l}{\pi \cdot d^2} \quad (1.1)$$

ρ is the specific electrical resistance and is given in the unit $\Omega \cdot mm^2/m$. Furthermore, l stand for the length of the conductor and A for the cross sectional area. Assuming round wires for coils the cross sectional area can be substituted by the wire gauge.

Magnetic field around a current-carrying conductor

Magnetic effects based on current-carrying conductors are used for technical applications because this effect can be controlled via the amperage, layout and number of conductors and the number of turns N. A straight conductor, which carries the current $\vec{I}$, creates a magnetic field with field lines of concentric circles in plains perpendicular to the conductor. In a given point P, the vector of the magnetic field strength of the magnetic field is called $\vec{H}$. Its direction is perpendicular to the current $\vec{I}$. The orientation of the field strength vector results from the motion of a right turning screw: as the screw turns in the direction of the field, it moves in the direction of the current $\vec{I}$ inside the conductor.

Figure 1.9 Rule by Lenz: magnetic field around a current carrying conductor

Another rule describing this relation is the right-hand rule: wrapping the right hand around the current-carrying conductor, the thumb points in the direction of the current, while the other fingers indicate the direction of the field lines. The value of the magnetic field strength $|\vec{H}|$ at point P is proportional to the current strength $\vec{I}$ inside the conductor and inversely proportional to the distance of point P to the conductor, as shown in Figure 1.9. Therefore, the field strength H is given by:

$$H = |\vec{H}| = \frac{I}{l} \quad \text{mit} \quad l = 2\pi \cdot r \quad (1.2)$$

where l is the length of the field line running through P. The unit of the magnetic field strength is ampere/meter. The structure of the field for increasing numbers of turns is shown in Figure 1.10.

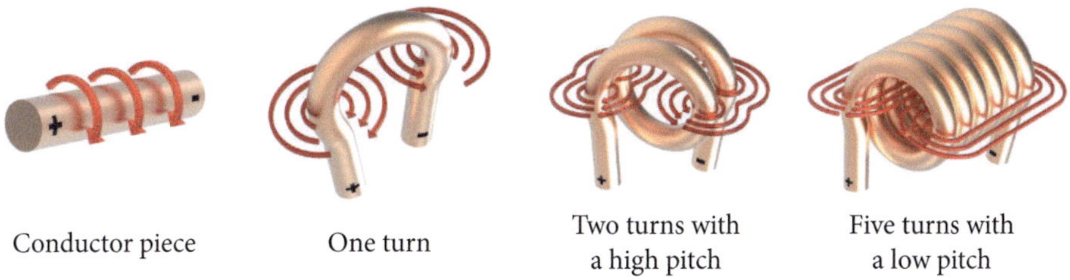

Figure 1.10 Structure of the magnetic field around a coil

Ampere's circuital law

For a closed loop (field line) in a magnetic field, the sum of the currents crossing this loop is called magneto motive force Θ of the loop:

$$\Theta = \sum_n I_n \tag{1.3}$$

When adding up individual currents the signs have to be taken into account, since simply adding the values will lead to an incorrect result. For a coil with N turns this means: for a coil without a core and carrying current $\vec{I}$, a random field line of the created magnetic field can be selected as a closed loop. The magneto motive force of this loop results in:

$$\Theta = N \cdot I \tag{1.4}$$

The magnetic field strength of this field line is

$$H_i = \frac{\Theta}{l_i} = \frac{N \cdot I}{l_i} \tag{1.5}$$

If the coil has a core made of ferromagnetic material, almost all of the field lines run through the iron core. This is because the magnetic resistance of air is significantly higher than for ferrimagnets. Accordingly, the field will choose the path of least magnetic resistance and the field lines will concentrate to the inside of the core. The magnetic resistance is also called reluctance and has the symbol RM. In this case the magnetic field strength in the core can be seen as almost homogenous and the average length of the field lines can be used to calculate the constant field strength:

$$H = \frac{\Theta}{l_m} = \frac{N \cdot I}{l_m} \tag{1.6}$$

The number of ampere-turns has established itself as a practical unit for the design and production of inductive components. This also represents the product of current and number of turns in a coil, and is therefore equivalent to the magneto-motive force.

Magnetic flux density or magnetic induction

A magnetic field is not fully described by the effect of the magnetic field strength. For a complete definition, another field parameter, magnetic flux $\vec{B}$, often referred to as magnetic induction, must be introduced.Its value is mainly defined by the properties of the medium in which the magnetic field is created. It runs in the same direction as the magnetic field strength and is defined by multiplying with a material dependent constant, the magnetic permeability μ.

$$\vec{B} = \mu \cdot \vec{H} \tag{1.7}$$

The permeability is the product of two values:

- dimensioned permeability in vacuum $\mu_{0:}$

$$\mu_0 = 4\pi \cdot 10^{-7} \text{ Vs/Am}$$

- dimensionless relative permeability of the material μ_r:

$$\mu = \mu_o \cdot \mu_r \tag{1.8}$$

The unit for magnetic flux density is Tesla with 1 T = 1 Vs/m^2. For non-magnetic materials the approximation is $\mu_r = 1$. For magnetic materials it is always $\mu_r > 1$. Ferromagnetic materials can in theory reach μr of up to 10^6, but are effectively in the area of 10^3 or 10^4. However, for these materials the relative permeability is not constant, instead it depends on both the properties of the Weiss domains in the materials and the current magneto motive force. This is based on the physical dependencies of atoms, such as Block walls between the domains or phenomena like the Barkhausen effect, a discrete magnetization of the material caused by lattice imperfections.

For illustration purposes the magnetic flux will be discussed using the example of a coil. It is initially assumed that the iron core in the coil is non-magnetic and that there is no current flowing in the coil. If the current is slowly increased in the coil starting from zero, a magnetic field develops, which extends predominantly in the iron core because of the path of least magnetic resistance. This field causes the initially randomly oriented Weiss domains to increasingly align in parallel to the field lines of the magnetic field. At first, this alignment is reversible, however it becomes increasingly irreversible as the field strength increases. By aligning these domains, also called elementary magnets, the magnetic flux density in ferromagnetic materials is significantly higher for higher field strengths than for smaller field strengths. This effect only continues until all of the molecular magnets are aligned. At this point saturation sets in and the magnetic flux density no longer increases. When reducing the coil's current after saturation, and therefore reducing field strength, only the reversible rotations of elementary magnets are changed so that a certain magnetic flux density remains, which is called remanence Br, even if the field created by the current-carrying coil is completely dismantled. The iron core has therefore become a permanent magnet. In order to demagnetise the core the direction of the coil current and consequently the orientation of the field must be reversed. The field strength of the opposing field necessary is called coercive field strength Hc of the material. With further increase of the current in the coil and the magnetic field in the ferromagnetic material, the process described above is repeated in the opposite direction until it reaches saturation again. When field strength B and flux density H are plotted (B-H diagram), the resulting loop is called magnetic hysteresis loop (Figure 1.11).

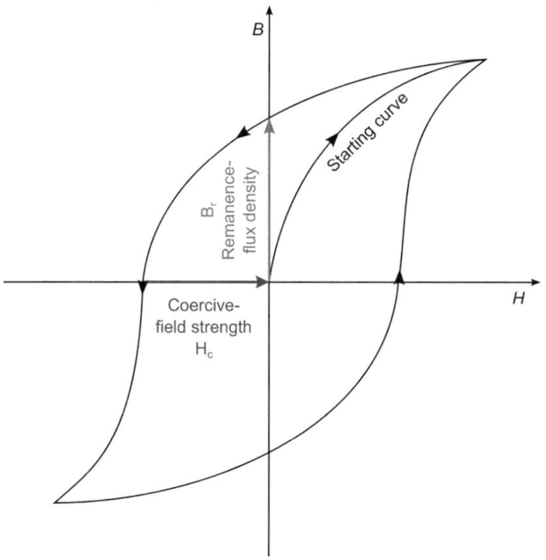

Figure 1.11 Physical effects of magnetization: hysteresis characteristics

For coils used in applications with alternating current, this process takes place at 50 Hz, meaning 50 times per second. The friction of molecular magnets aligning inside the material causes losses, which are proportional to the core's volume and the area of the hysteresis loop. For electrical machines and equipment (e.g. motors and transformers), in which the field lines run through sheeted metal structures these losses are undesirable for two reasons: firstly, because the efficiency decreases and secondly, because the components heat up. To counter this effect electromagnets, transformers and electric motors use magnetically soft materials, which have the smallest possible remanence and coercive field strength. This applies to so-called dynamo sheet metal (Fe with 3.75 % Si).

Permanent magnets, however, must have high remanence and high coercive field strength to protect against unwanted demagnetization by electrical stray fields. This is ensured by the use of so-called hard magnetic materials for permanent magnets. The hardest magnetic materials, whose remanence is almost as high as their saturation flux density and whose coercive field strength almost reaches the saturation field strength are referred to as rectangular-hysteresis ferrites.

Magnetic flux

The magnetic flux Φ is the unity of field lines emerging from one area and the field lines entering the opposing area. Here, A is the area occupied by the magnetic field, meaning the area where the magnetic flux spreads in the air gap. Therefore, flux density measures the flux in a certain area. Thus, flux equals flux density times area. For a homogenous magnetic field, the definition for magnetic flux is defined by pole area A, which is arranged perpendicularly to the flux direction:

$$\Phi = B \cdot A \tag{1.9}$$

The unit for magnetic flux is Weber (volt-second) 1 Vs = 1 Weber = 1 Wb.

Analogies of mechanics and electrodynamics

For a better comprehension of electrical and magnetic field phenomena analogies with mechanics are often used. A selection of such analogies is compiled in Table 1.2.

Table 1.2 Analogies from mechanical and electrical engineering according to [Alt-08]

Mechanical Unit		Electric analogy 1		Electric analogy 2	
Force	F	Voltage	U	Current	I
Velocity	v	Current	i	Voltage	u
Path	s	Charge	Q	Magnetic flux	Φ
Force pulse	Δp	Magnetic flux	Φ	Charge	Q
Mass	m	Inductance	L	Capacity	C
Velocity	$v = \Delta s/\Delta t$	Current	$v = \Delta Q/\Delta t$	Voltage	$v = \Delta\Phi/\Delta t$
Force of inertia	$F = \frac{\Delta p}{\Delta t} = m \cdot \frac{\Delta v}{\Delta t}$	Voltage	$U = \frac{\Delta\Phi}{\Delta t} = L \cdot \frac{\Delta i}{\Delta t}$	Current	$I = \frac{\Delta Q}{\Delta t} = C \cdot \frac{\Delta u}{\Delta t}$
Energy	$\Delta W = F \cdot \Delta s$	Energy	$\Delta W = U \cdot \Delta Q$	Energy	$\Delta W = I \cdot \Delta\Phi$
Power	$P = F \cdot v$	Power	$P = U \cdot I$	Power	$P = I \cdot U$
Spring rate	$D = F/\Delta s$	Capacity	$C = \varepsilon \cdot A/d$	Inductance	$L = w^2 \cdot \lambda$
Movement	$p = m \cdot v$	Magnetic flux	$\Phi = L \cdot i$	Electric charge	$Q = C \cdot u$
Kinetic energy	$E = \frac{1}{2} m \cdot v^2$	Magnetic energy	$E = \frac{1}{2} L \cdot i^2$	Electric energy	$E = \frac{1}{2} C \cdot u^2$

Lorentz force

If an electrical charge q is moving through an electromagnetic field with a defined velocity v it is affected by the so-called Lorentz force F_L. Mathematically this relationship for a point charge q can be described by:

$$F_L = q \cdot (\vec{v} \times \vec{B}) \tag{1.10}$$

In winding technology and electric motors this is practically relevant to extend the phenomenon of a current-carrying conductor with the length l. It applies in the following way: if an electrical conductor is orientated perpendicularly to a surrounding magnetic field and carries a current and external force can be observed. Assuming that the conductor's orientation and the magnetic field are perpendicular, the following applies:

$$F_L = I \cdot l \cdot B | \vec{l} \perp \vec{B} \tag{1.11}$$

This is often exemplified by using a conductor swing. If an electrically contacted conductor loop is located between the sides of a magnetic circle the conductor swing is deflected towards the sides by the Lorentz force. Further and more detailed explanations can be found within the extensive literature on the nature of the Lorentz force.

Law of induction

When a constant magnetic flux penetrates a conductor loop with N turns, which is not carrying a current and is at rest, no forces are exerted on the electrons inside the conductor. However, if the magnetic flux changes, a current *i(t)* can be detected in the conductor loop, the size of which is proportional to the speed of change of the magnetic flux. The cause of this current must be a source of voltage U_q induced by the change in flux:

$$u_q = N \cdot \frac{d\Phi}{dt} \tag{1.12}$$

Coil inductivity

To explain a coil's self-inductance, an electrical coil with its conductors wound around a ring-shaped core made of non-magnetic synthetic material (ring coil) is examined. Since the non-magnetic synthetic material has almost the same permeability as a vacuum or air it may also be called air coil. The number of turns of the coil is *N*, the coil's cross sectional area is *A* (cross sectional area of the core) and the average length of field lines is *l*. For a current changing over time *i(t)* the flow

$$\Theta(t) = N \cdot i(t) = H(t) \cdot l \tag{1.13}$$

is observed through the coil. Changing the current over time changes the flow over time. The magnetic flux is therefore

$$\Phi(t) = B(t) \cdot A = \mu_0 \cdot H(t) \cdot A = \frac{\mu_0 \cdot N \cdot i(t) \cdot A}{l}. \tag{1.14}$$

According to the law of induction the changing magnetic flux causes the induction of a voltage, which in turn causes a current to flow in the coil. This is called self-induction. The induced voltage is often 1 *nH* per

$$u(t) = N \cdot \frac{d\Phi}{dt} = \frac{\mu_0 \cdot N^2 \cdot A}{l} \cdot \frac{di(t)}{dt}. \tag{1.15}$$

It is therefore proportional to the coil current's derivation over time. The factor of this proportionality depends on the number of turns N, the permeability μ_0 and the dimensions of the coil. This factor, characteristic for self-inductance, is referred to as inductivity L:

$$L = \frac{\mu_0 \cdot N^2 \cdot A}{l}. \tag{1.16}$$

Inductivity, however, also depends on the design of the coil. For ring coils, for example, the length and diameter of the core are decisive. A distinction is made between the outer radius of the core R, the inner radius of the core r and the resulting core's width or depth b. As an approximation the following equation can be used:

$$L = \frac{\mu_0 \cdot \mu_r \cdot N^2 \cdot b}{2\pi} \ln \frac{R}{r}. \tag{1.17}$$

For air gaps in the shape of cylinders, which have at least a 0.6-fold length to the radius r_w, the following approximation is valid:

$$L = \frac{\mu_0 \cdot N^2 \cdot A}{l + 2 \cdot r_w/2,2}. \tag{1.18}$$

For practical purposes other characteristics like the induction constant A_L, are often suitable because pre-manufactured cores are used. The unit of this characteristic value is often nH per square turn, which complies with reciprocal of the magnetic resistance. The inductivity is defined as follows, but is only valid for the linear part of the magnetic flux without saturation:

$$L = A_L \cdot N^2. \tag{1.19}$$

Consequently, the following correlation exists for current and voltage of a coil

$$u(t) = L \cdot \frac{di(t)}{dt}. \tag{1.20}$$

The inductivity's unit is 1 Henry = 1 Vs/A = 1 Ωs (1 H).

An elongated rectangle is used as a circuit symbol for coils. However, in contrast to the circuit symbol for an electrical resistance it is filled in black. A real coil is always made from conductor wire, so that it always has an electrical resistance. Often a stylised coil is used as the circuit symbol for a real coil or, in the equivalent circuit diagram, a series connection of an ideal coil and an electrical resistance. The voltage drop in a real coil is therefore:

$$u(t) = u_R(t) + u_L(t) = R \cdot i(t) + L \cdot \frac{di(t)}{dt}. \tag{1.21}$$

Electrical power losses in coils

The power losses P_{Cu} in a coil depend primarily on the electrical resistance and can therefore be described as [Wic-00]:

$$P_{Cu} = I^2 \cdot R_{Cu} = J^2 \cdot A_{Cu}^2 \cdot \frac{\rho_{Cu} \cdot l_{Cu}}{A_{Cu}} = J \cdot A_{Cu} \cdot \rho_{Cu} \cdot n \cdot l_m$$

$$P_{Cu} = J \cdot I \cdot n \cdot \rho_{Cu} \cdot l_m = J \cdot \Theta \cdot \rho_{Cu} \cdot l_m \tag{1.22}$$

Here, the current density J describes the current divided by the conductor's cross sectional area A_{Cu}. The value l_m describes the average length of the conductor per turn, which gives the total conductor length when combined with the number of turns. However, this formula applies for flux density

$$P_{Cu} = n \cdot I = J \cdot A_{Cu} \cdot n = J \cdot k_{Cu} \cdot A_W \quad \text{bzw.} \quad A_W = \frac{\Theta}{J \cdot k_{Cu}}. \tag{1.23}$$

In the last expression further parameters are introduced. k_{Cu} is a key characteristic in winding technology and describes the ratio between the area of the copper wire and the available winding area. The cross sectional area of the complete winding including gaps is described by the parameter A_w. The last unknown parameter in the power loss equation is the average wire length l_m. For a non-circular coil core it can be expressed with the following equation, where b and t are the bobbin's width and depth, and b_w describes the winding width:

$$l_m = 2 \cdot (b + t + 2 \cdot b_w) = 2 \cdot (b + t + 2 \cdot \frac{A_W}{h_W}) = 2 \cdot (b + t + 2 \cdot \frac{2 \cdot \Theta}{h_W \cdot J \cdot k_{Cu}}). \tag{1.24}$$

Using this relationship the power loss can be expressed as:

$$P_{Cu} = J \cdot \Theta \cdot \rho_{Cu} \cdot 2 \cdot (b + t + \frac{2 \cdot \Theta}{h_W \cdot J \cdot k_{Cu}}). \quad (1.25)$$

Electrical losses in coils therefore depend on two general aspects. Outside the brackets are current density, magnetic flux and electrical conductivity. All of these quantities correspond to boundary conditions of the application, like current and number of turns as well as the choice of conductor material and the wire cross sectional area depending on the current density. The following term is primarily defined by the coil's shape and the fill factor, which is why it can be more strongly affected by manufacturing and winding technology.

Magnetic energy of a coil (air coil)

The magnetic energy of a coil equals the electrical work while building the magnetic field:

$$dW_m = dW_{el} = u(t) \cdot i(t) \cdot dt = L \cdot \frac{di}{dt} \cdot i(t) \cdot dt = L \cdot i \cdot di. \quad (1.26)$$

Thus

$$W_m = L \cdot \int i \cdot di = \frac{1}{2} \cdot L \cdot I^2. \quad (1.27)$$

Calculating the magnetic energy for iron coils, however, is more complex, because the coil's inductance L is significantly larger than for air coils, but it depends on the current *i(t)* like the permeability μ.

Magnetic circuits

A closed field line in a magnetic field can also be called magnetic circuit, due to the analogy to an electrical circuit. The source of any electrical current, the voltage U_q, is compared to the magneto motive force Θ seen as the source of the magnetic field and is therefore also referred to as magnetic potential difference. The electrical conductivity $\kappa = 1/\rho$ equates to the permeability $\mu = \mu_0 \cdot \mu_r$ in magnetic circuits, so that the electrical resistance $R = L/(\kappa \cdot A)$ compares to the magnetic resistance in the core $R_M = l_m/(\mu \cdot A)$.

The effect of the electrical voltage in an electrical circuit is an electrical current $I = U_q/R$, while the effect of the magnetic voltage is the magnetic flux $\Phi = \Theta/R_M$. Consequently, analogue to the current density $S = dI/dA$ in electrical circuits, there is magnetic flux density in magnetic circuits

$$B = \frac{d\Theta}{dA}. \quad (1.28)$$

Simple magnetic circuit

A coil with a closed iron core can be referred to as a simple magnetic circuit. Knowing that the core's cross sectional area is constant, the current I required to create a desired magnetic flux Φ depending on the average field line length l_m and the number of turns can be determined.

Approach: The required flux density in the iron core is

$$B_{Fe} = \frac{\Phi}{A}. \tag{1.29}$$

Therefore, the required field strength H_{erf} can be gathered from the core material's magnetization characteristic provided in a diagram by the manufacturer of the core material (Figure 1.11), if not stated otherwise.

$$B_{Fe} = f(H_{Fe}). \tag{1.30}$$

The required magneto motive force results to

$$\Theta_{erf} = H_{erf} \cdot l_m \tag{1.31}$$

and the required coil current to

$$I_{erf} = \frac{\Theta_{erf}}{N}. \tag{1.32}$$

Magnetic circuit with air gap

If the previously discussed simple magnetic circuit's iron core contains a tight air gap, which has a width b that is a lot smaller than the average length of the field lines in the iron core, it can be assumed that the magnetic field still flows through the iron core, and that stray losses will be negligible. The magnetic flux density is therefore the same in the iron as in the air gap and is the result of the quotient of the magnetic flux and the core's cross sectional area:

$$B_L = B_{Fe} = \frac{\Phi}{A}. \tag{1.33}$$

The core material's magnetization characteristic

$$B_{Fe} = f(H_{Fe}) \tag{1.34}$$

gives the required field strength H_{Fe} in the iron core. The field strength in the air gap is

$$H_L = \frac{B_L}{\mu_0}. \tag{1.35}$$

Ampere's circuit law

$$\int \vec{H} \cdot d\vec{l} = \Theta \tag{1.36}$$

becomes

$$\Theta = H_{Fe} \cdot L_{Fe} + H_L \cdot d = H_{eff} \cdot L_m \tag{1.37}$$

to arrange the core and air gap to give the most effective field strength H_{eff}.

When plotting the flux density over the effective field strength, the magnetization curve for the core with air gap is flatter and more linear than the curve for a closed core, and is referred to as sheared magnetization curve of the arrangement.

Energy of magnetic fields

For a current-carrying coil with closed iron core the energy stored in the magnetic field of the iron core at the desired flux density is equal to

$$W_{Fe} = V_{Fe} \cdot \int H_{Fe} \cdot dB_{Fe}, \tag{1.38}$$

where V_{Fe} is the volume of the iron core.

Due to the linear relation between magnetic flux density and magnetic field strength, the energy stored in the air gap for an iron core with air gap is equal to

$$W_L = \frac{B_L \cdot H_L \cdot V_L}{2} = \frac{B_L^2 \cdot V_L}{2 \cdot \mu_0}. \tag{1.39}$$

For a tight air gap in the iron core, the magnetic flux density in the air gap is the same as in the iron. Because of the much smaller permeability of air, the magnetic field strength in the air gap is substantially bigger than in the iron core.

As a result, at equal flux density the iron core with air gap stores more energy than a closed iron core. However, a higher coil current and therefore higher supply of electrical energy is required to create the same flux density as in a comparable circuit with closed iron core.

Relevant physical effects for the operation of inductances

So-called eddy currents are a relevant effect for electric motors. As introduced in the inductance section, a magnetic field changing over time causes an inductance in an extended electrical conductor. The same applies if a current-carrying electrical conductor is moved through a constant magnetic field. Generally speaking, a voltage is induced, which, depending on the magnetic flux density and the electrical resistance of the object, results in eddy currents. The term for eddy currents originates in local changes of the magnetic field towards an inhomogeneous distribution. To contain the negative effects of eddy currents without influencing the magnetic properties, electric motors therefore utilise individually insulated sheeted structure of the coil's core to keep the electrically conducting layers thin (Figure 1.12). These thin layers in turn result in higher electrical resistances, which reduce eddy currents at equal inductance.

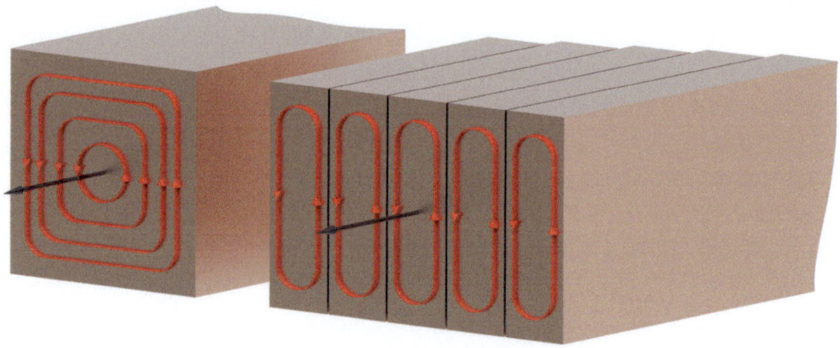

Figure 1.12 Influence of a lamination stack on the eddy currents

The induced current creates a magnetic field around the local conductor according to Lenz' law, which opposes the exciter field. At the center of the conductor the opposing field is stronger than at the edge. For an inductance with high frequencies the outer magnetic field can no longer penetrate the conductor's core with the induced voltage because of this so-called current suppression. Consequently, the magnetic field's depth of penetration into a conductor changes with its frequency. This phenomenon is called skin effect, since high-frequency applications conduct current only in the outer layers. This is why only smaller wire gauges are generally suitable for high-frequency coils. Increasing the diameter would not increase the effective conductor cross sectional area. Stranded wire is often used for such applications (Section 2.1.3). Table 1.3 shows typical penetration depths according to the skin effect for copper conductors.

Table 1.3 Skin-depth of current depending on the excitation frequency

Low-frequency range (LF)		**Middle-frequency range (MF)**		**High-frequency range (HF)**	
Frequency	Skin depth	Frequency	Skin depth	Frequency	Skin depth
500 Hz	2.97 mm	500 kHz	0.0938 mm	500 Mhz	2.97 μm
5 kHz	0.938 mm	5 MHz	29.7 μm	5 GHz	0.938 μm
50 kHz	0.297 mm	50 MHz	9.38 μm	50 GHz	0.297 μm

A similar relationship for the interaction with magnetic fields may also occur for adjacent conductors. The suppression of current between two adjacent conductors is triggered by magnetic flux leakage. Accordingly, as in the skin effect, the conductor's cross sectional area is only partially used to carry the current. This so-called proximity effect depends on the distance between the conductors, the conductor's magnetic conductivity and the magnetic field's frequency. Due to the dependency on the distance, a symmetrical structure is important for compact windings.

In electrodynamics there are further effects, however, these are less relevant for the operation of inductive components. Typically, these effects are used for sensory tasks to gather information. In this context they are exemplified by the Hall Effect and the magneto resistive effect. The Hall Effect describes the occurrence of a voltage in a current-carrying inductor, when, as opposed to the effect of inductance, it is located in a stationary magnetic field, meaning that it is not changing over time. Here, the voltage is perpendicular to the direction of the current or the magnetic field and is referred to as Hall voltage. This is caused by the joint occurrence of electrical and magnetic fields. The stationary magnetic field causes an electrical field perpendicular to the actual direction of the current. As a result, a voltage can be tapped from the conductor in this direction, which is proportional to the strength of the magnetic field. This is why this effect is used to measure magnetic fields with so-called Hall probes. The Hall voltage U_H results from:

$$U_H = A_H \cdot \frac{I \cdot B}{d}. \tag{1.40}$$

A_H is a material constant, also referred to as Hall constant, and d the width of the probe. The magneto resistive effect is a collection of multiple operating principles, which cause a change in the electrical resistance of a conductor by applying an external magnetic field. These effects have their roots on the microscopic level of materials and can generally be described by the influence of the type and magnitude of the material's elementary magnets on the electrical conductivity. This effect is utilised in both magnetic sensor technology and storage technology. In magnet electronics it is possible to write information into storage cells by applying a magnetic field, and read out by determining the electrical resistance.

Resistance and inductance of a coil are relevant to winding technology. In practice, ampere-turns play a major role in coil design. Warranting the properties defined by the product's designer is a key task in coil winding technology, although the coil geometry may change due to the winding process.

1.2.3 Coil types

Coils can be classified by the type of application, the coil's structure and the manufacturing properties. Alternatively, coils can be divided into coils without a magnetic core, or air coils, and coils with bobbins. When differentiating by structure, cylindrical, toroidal and disk coils can be distinguished, while for motors it is either concentrated or distributed windings. These are shown in Figure 1.13 and 1.14.

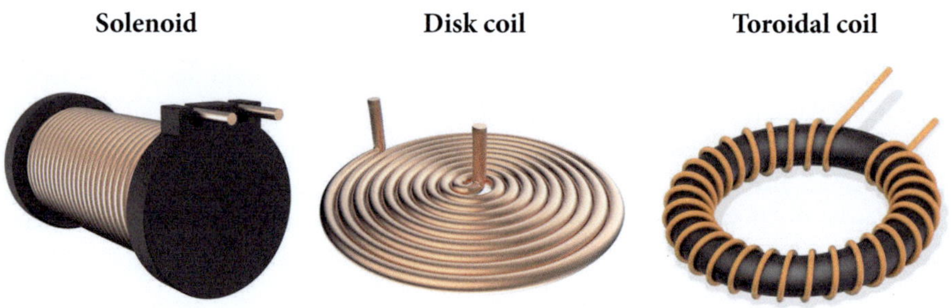

Figure 1.13 Different shapes of wound coils

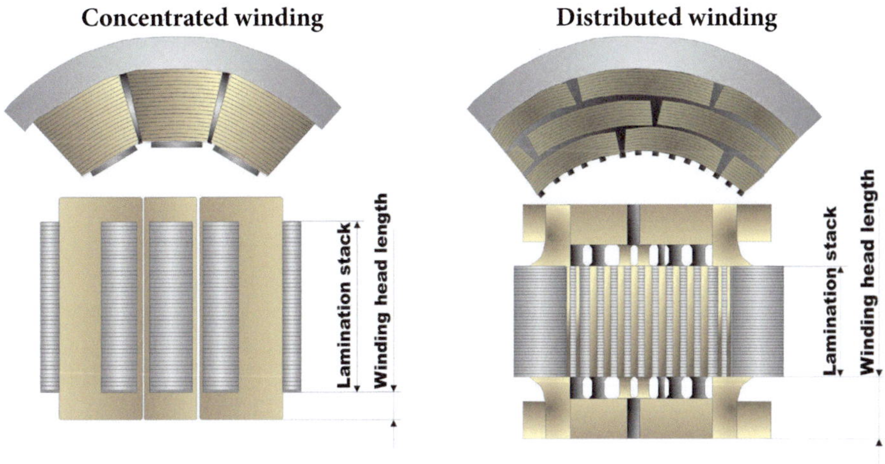

Figure 1.14 Different types of motor windings

The following Figure 1.15 classifies cylindrical coils according to their areas of applications.

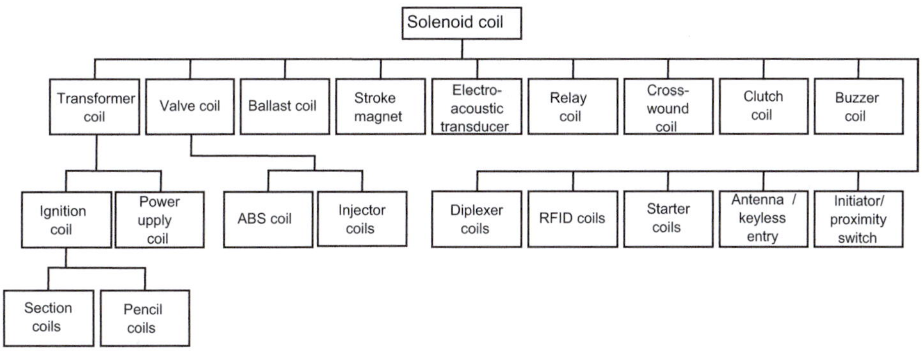

Figure 1.15 Characterization of different applications for solenoid coils

Figure 1.16 shows the classification of disk coils as well as concentrated windings for their areas of applications. Distributed windings are only used for sheeted stators and rotors. Toroidal coils are typically used for transformers.

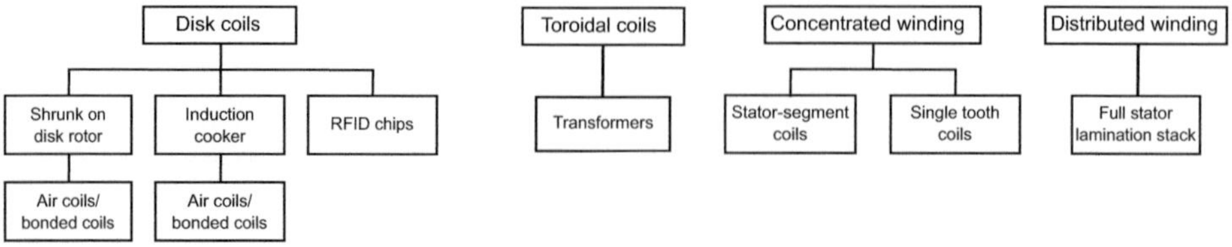

Figure 1.16 Characterization of different applications for the other coil shapes

In the following, coils are clustered by their application including an introduction to each function and application respectively. The first category describes the use of the magnetic field with a core (magnetic flux) for electromechanical energy transformation or actuators, such as motors, generators or solenoids. A special application of solenoids is the area of acoustic transducers and actuators for speakers or microphones. Compensation of reactive power in power supply equipment is an application of restrictors. Another type exists to transform different voltages through a closed magnetic circuit with different number of turns on primary and secondary sides of transformers, which can be used for ignition coils as well. Typical values like wire gauge, operating voltage and number of turns for different types of coils are shown in Figure 1.17.

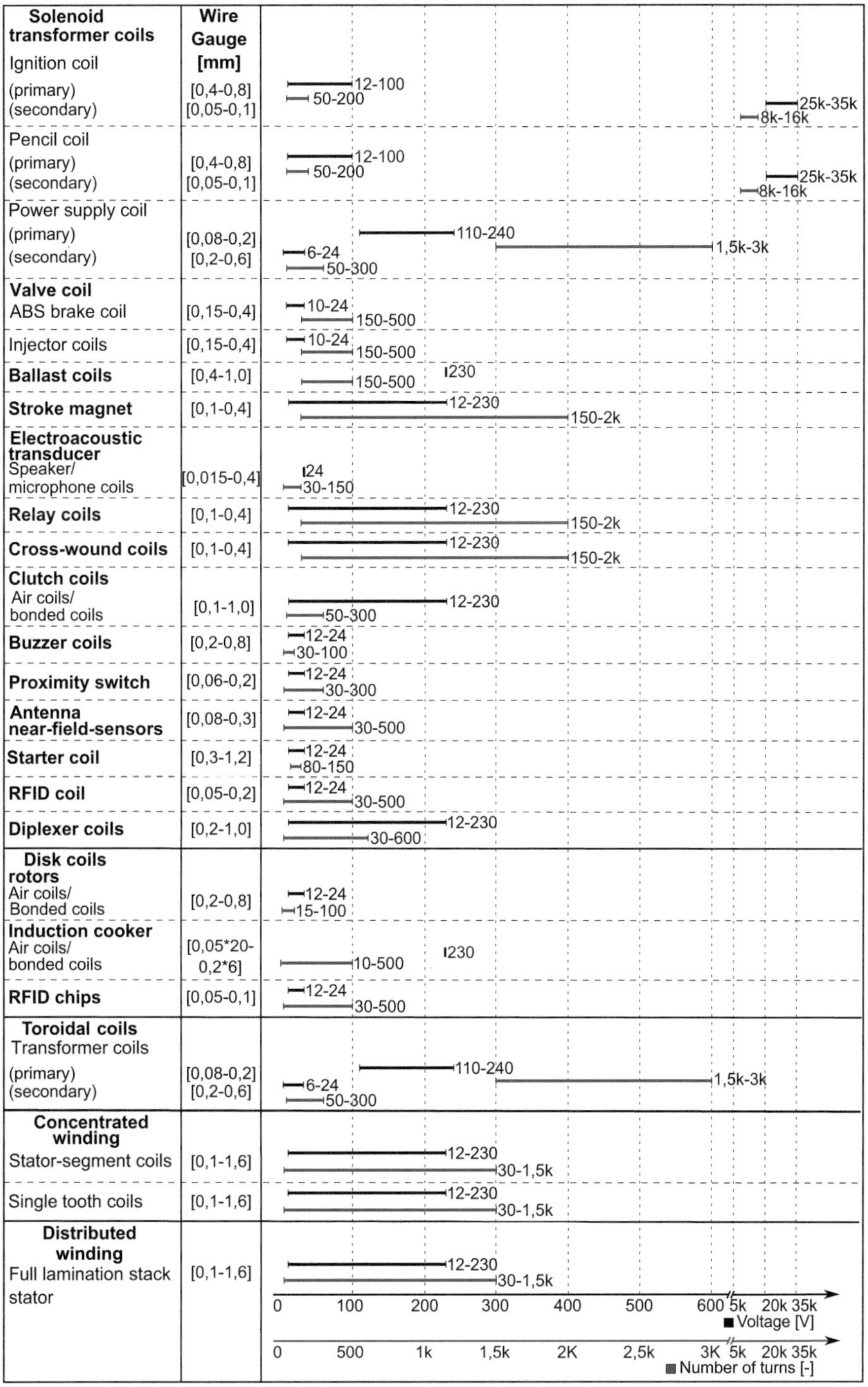

Figure 1.17 Overview of typical voltages and number of turns per coil application

The second category of possible inductance applications uses the rise and fall characteristics of voltage and current, and their specific time coefficients for signal processing. Using these coefficients, oscillating circuits and filters (high-pass, low-pass and band-pass) are adjusted to the desired frequencies or phase. The third category covers the use of magnetic fields without a core, meaning electromagnetic waves. They can be applied to transfer information in telecommunications (high-frequency antennas), to transmit electrical power through inductive charging using transformers, or to detect objects using changes in the field and the proximity effect in sensor technology. A special application of these fields is based on the interaction with charged particles, as in guiding individual particles in particle accelerators, like the Large Hadron Collider at CERN. Furthermore, non-inductive coils can be used as resistances in the area of resistive passive components by winding chrome nickel constantan wires. These are used as heating resistances, power resistances, or as motor protection with bi-metals. Non-electrical applications such as coils made of welding wire or fishing thread should be mentioned, however, they are not emphasised because of their non-electrical application.

Labels for various coils and corresponding circuit symbols are provided by the standard EN-60617-4, but are not listed due to their large numbers. For winding technology the general requirement from various applications is to manufacture a coil with the appropriate electrical resistance and a tolerance of ± 5 %.

A coil is a component that is customised to the respective application, with its geometry primarily adapted to the available assembly space and its structure adapted to the function.

1.2.4 Introduction to bobbins

Figure 1.18 A coil bobbin optimised for winding processes

In comparison to bobbins of the past, which were simply used as winding carriers, bobbins today are highly optimised components, which are modified for each application. Bobbins play a vital role for the electromagnetic properties of coils. The bobbin geometry therefore also plays an important role for design and industrial manufacturing of coils. The original function of the bobbins and winding carriers was to carry the winding, to give it spatial limits and to provide a basic electrical insulation. However, the insulation of the bobbin varies greatly depending on the application. In addition to electrical boundary conditions, such as voltage and insulation strength, ambient conditions, including humidity and media, play a significant role. An introduction to the different insulation properties of bobbin materials is given in Section 1.2.5. Besides manufacturing from a certain material, for example injection moulding, other processes can also be used to insulate, for instance spraying or coating the basic body (for example with powder coating, oxidation or anodization), or by assembling further components (for example insulation paper or foil) as used for motor technology. One typical application would be the use of insulation paper for the pocket insulation and plastic end caps, or end plates for the tooth head insulation. Wound up bodiless wire coils (air coils) have almost no solid structure, making further processing operations difficult. However, the most important and initial aspect of bobbins is the stabilization and spatial limitation of the coil. At the beginning of the 20th century bobbins were made from cardboard or Bakelite, an early synthetic material. In a largely manual fabrication, the coil ends where fixed with bee wax or thread after winding. The coil's contacts could then later only be contacted by hand, for example by soldering, bolting or crimping. As a result of the growing requirements for further industrial processing of coils bobbins were assigned with more functionality. Due to the limited choice of suitable materials at the time, like laminated paper or thermosetting materials, such as Bakelite, the limits of performance and shapes were reached quickly. Adopting further functionalities of the bobbin was mainly enabled by further development of thermosetting synthetics. The performance of bobbins today is characterised by the following physical and geometric properties:

- temperature resistance
- electromagnetic properties
- insulation properties
- shape stability
- ways of miniaturization
- abilities of automation.

In the light of today's high requirements for energy efficiency of mechatronic components, power density is a priority. Further functionalities assigned to the bobbin simplify the electrical contracting and later processing. Modern injection moulding technology enables the integration of bearing seats, valve stem seals, various ways of contacting, or synthetic joints. Pins, which are differentiated into park pins, auxiliary pins and wrapping pins, are commonly used (Figure 1.4). The park pin is used to fixate the wire on the winding tool between winding processes. The auxiliary pin helps guide the wire on the bobbin or winding tool, while the wrapping pin is needed to terminate the wire. Bobbins usually have a winding ground and lateral limits of the winding space, which are also referred to as bobbin flange. Ultimately, in the course of automation a number of functionalities were assigned to the bobbin. A variety of bobbin geometries may be utilised depending on the application. Bobbins with angular cross sections are mainly used for voltage converters and coils with round cross sections were developed for mechatronic systems. The most common bobbin types are described in the following paragraphs. Exceptions are the so-called bodiless coils. These are usually divided into two categories (Figure 1.19).

Figure 1.19 Examples for air coils, bonded air coils and transformer coils

Bodiless air or field coils

These coils are wound as dimensionally unstable coils (Figure 1.19). A winding device provides the limitation of the winding space in the shape of the winding ground and winding flanges. After the winding process the turns are fixated, for example with tape, for removal from the winding tool and for further processing. Another form of bodiless air coils are the single coils for distributed motor windings, which are neither wrapped nor taped. After winding they are transferred to another tool, which can accommodate multiple air coils, to manufacture stator windings, e.g. for asynchronous motors and field coils (for power tools). Subsequent to windings and transfer, all of the air coils are inserted into the pockets of the stator using the described carrier tool. Only the winding heads may be taped, shaped or optionally bonded with resin after this so-called insert process.

Bodiless bonding coil

With the exception of hot-air bonded coils (Section 3.4.3), these coils are not wound as dimensionally stable coils (Figure 1.19). Similar to bodiless air coils, a winding tool provides the limitations of the winding space. To achieve a winding with dimensional stability the copper wire needs to be coated thermoplastically (self-bond enamel), which barely contributes to the wire's insulation. This coating is re-softened and then hardened (baking) during or immediately after winding before the coil is taken from the winding tool. A distinction is made between three types of baking processes; these are covered in more detail in Section 3.4.3. The bobbins, also referred to as winding carriers, usually have a winding ground and bobbin flanges. The different bobbin geometries result from the various applications. Bodiless baked coils are also used for iron-less stators in DC motors. Criteria, such as the coil's assembly room, the electrical insulation, the mechanical protection of the winding as well as manufacturing and economical factors must be considered. Bobbins can also consist of several sub-assemblies. Individual end caps for full lamination package stators can also be referred to as parts of the bobbin. In the detailed remarks about bobbins a number of things, such as the design of the winding space, need to be considered. Systemic winding carriers, which mainly include other functional sub-assemblies, are referred to as winding tools, not bobbins. These include, for example, coated sheet metal packages of stators, single teeth or rotors. Since coils and windings are the key part of mechatronic systems, one tries, based on the different production technologies, to divide the value added during the manufacturing process. In this respect a coil on a bobbin will only be processed further, if it meets the requirements in terms of quality. Frequently this is not the case for functional sub-assemblies. Coated stators, for example, are wound directly and the value added is not interrupted, so that a reject part results in the rejection of a complete sub-assembly. Due to the number of individual coils or phases, these winding carriers are also referred to as winding carrier systems rather than bobbins. Stators with pockets on the inside and outside and ring core coils also count as winding carrier systems. A very special type of bobbin is a foil, which acts as winding carrier. It is used especially for acoustic transformers, such as speaker or microphone coils. These foils made from aluminum or Nomex insulation material, are usually shaped into the shape of a cylindrical sleeve upon which the self-bond enameled wire is wound.

Transformer bobbins

EI and UI transformer bobbins (Figure 1.19) are used in power supplies and shown in Figure 1.20. The letter combination is a typical expression for the shape of the core and describes the segments of the structure.

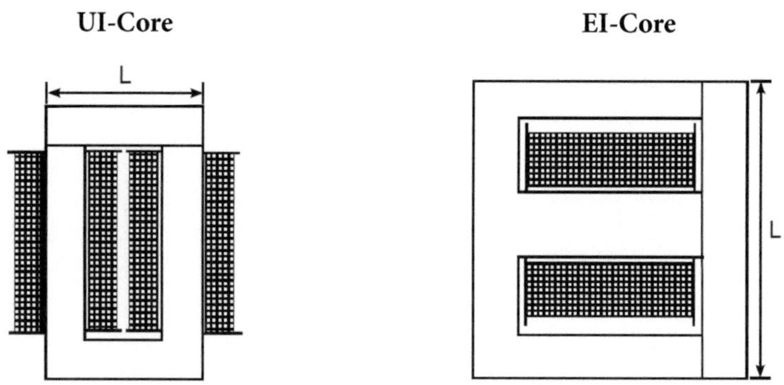

Figure 1.20 Different transformer core shapes according to [Reh-03]

Power supply units transform the relation between current and voltage. The most common form of power supplies are ballast units for use in low-voltage devices, such as phones or electrical shavers. If designed as EI transformer, the packaged sheets have separate shapes of Es and Is. This type of transformer is mostly used for low power transformers. The labelling UI means that the element transmitting the magnetic field is a packed iron element in shapes of Us and Is. After assembling two wound bobbins with a primary winding on one side and the secondary winding on the other side, the U profile is mounted on the I bar to bridge the magnetic fields. These sheet metal packages and can be clamped, bonded, screwed, riveted, or welded. Bobbins used in UI transformers share a winding space, which carries the primary and secondary winding. After assembling the wound bobbin the E-core is mounted to the I-core to bridge the magnetic fields. Following this, the I-core is fixated with the named processes. The bobbins used in EI transformers usually have two winding spaces separated by a bar, which carry the primary winding on one side and the secondary winding on the other side. Transformer bobbins usually have a row of contacting pins. Multiple contacts on each side in the shape of axial or radial contacting pins enable different voltage levels. In order to integrate those contact points the bent pins are moulded into the synthetic material to enable contacting from both sides, meaning contacting with wire on one side, and contacting with conducting paths of a polychlorinated biphenyl (PCB) circuit board on the other side. Primary and secondary winding are often spatially separated in both types of transformers. The spatial separation is achieved axially in EI bobbins and radially in EI bobbins. This is useful, because the higher primary voltage is usually spatially separated from the secondary voltage in ballast units and therefore meets a higher safety standard. The radial insulation in UI transformer coils usually consists of an adhesive, electrically insulating tape material or a strip of pressboard.

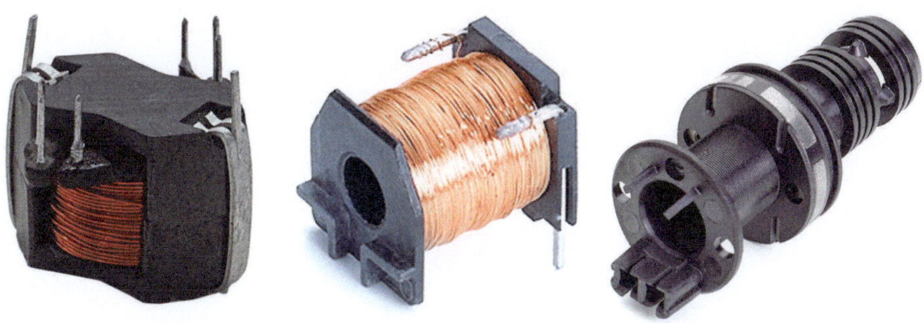

Pot core bobbin **Relay bobbin** **Valve bobbin**

Figure 1.21 Examples of a pot core bobbin, a relay bobbin and a valve bobbin

R_M bobbins and pot core bobbins

In contrast to typical smaller transformers, which are built with sheeted metal for magnetic flux, a ferrite core or ferrite pot core housing is used in these bobbins (Figure 1.21). These ferrite objects create the optimal magnetic saturation, since almost all of the object's lateral and front surface can be enclosed with ferromagnetic material. In addition, the frequency noise typically emitted by small transformers can be avoided with the help of this ferrite component. Similar to small transformers these bobbins have moulded, mostly bent contacts. R_M bobbins (R_M for magnetic resistance of the core material) originated in telecommunications and are used to transform small voltages with sometimes high frequencies. They were formerly used as pulse amplifiers.

Relay bobbins

The most common type of bobbin is the relay bobbin (Figure 1.21). Relays are electromagnetic switches, which close or open one or multiple outputs of electrical circuits depending on the incoming current. Often these are used in staircase lighting, contactors, automobile manufacturing (e.g. indicators) or motor protection switches. The winding ground of relay bobbins is usually round. Relay bobbins often have two contacting pins. To enable the use in combination with a circuit board, relay bobbin pins often go through the bobbin flange. Using this pin design the electrical connection of copper wire and contacting pin, as well as of contact area and circuit board, can be spatially separated. When using such continuous contacting pins the ends, which connect to the copper wire, are bent over using soldering or arc welding after contacting. This enables a highly space-saving design and reduces the stress on the wire leading to the winding, especially when using thin wires. Thus the vibrational resistance is improved as well. If one-sided inserted contacting pins are used, there is a dependency on the electrical connection type from copper wire to contacting pin, and from contacting pin to the interface with the actual connection.

Valve bobbins

Just like relay bobbins, valve bobbins (Figure 1.21) are also used in household, industry and automobile manufacturing. In addition to fixating the winding, the bobbins are often used for tasks of fluid mechanics. Common applications are gas or oil heating systems, water taps or automobile manufacturing. In terms of industrial applications, valve coils are used for control purposes, such as pneumatic or hydraulic devices. However, the main field of application for valve coils is automobile manufacturing. Valve coils are also found in fuel supplies and controls, seat adjustments including ventilation and massage functions or oil flow control in automatic gearboxes. Apart from a few exceptions, valve bobbins are designed as round winding carriers. The operating voltages usually do not exceed medium voltages. Due to the growing improvements in injection moulding technologies for synthetic materials, the bobbin not only fixates the winding, but can often also include flow passages. Sealing surfaces, core guidances and tube connections included in the mould turn valve bobbins into efficient functional systems. More and more functional standard valve components are integrated into the bobbin. Using only a few additional parts, such as springs, cores and membranes as well as inference cores or inference housings, a fully functional valve can be created. Valve coils usually only have one turn, which is designed for a certain voltage area. After inserting all of the necessary parts into the valve bobbin, the sub-assembly is often casted or over moulded.

Section bobbin **Pencil coil bobbin** **Cross-wound coil bobbin**

Figure 1.22 Examples of a section bobbin, a pencil coil bobbin and a cross-wound coil bobbin

Section bobbins

Section bobbins are designed for operating under high voltages, including ignition coils. Depending on the installation location section coils can be round, quadratic or rectangular (Figure 1.22). Applications are mainly ignition systems for heating equipment or in automobile manufacturing. A key characteristic of section coils are the flanges between the winding sections. Section bobbins are actually the secondary part of transformers, although for very high voltages. While in small transformers the primary voltage is usually transformed to a lower secondary voltage, in ignition transformers the secondary voltage is stepped up. The primary part of the ignition systems is usually located within the inner diameter of the section bobbin. Voltages of up to 30 kV can be created. Since handling such high voltages requires a special insulation standard for the winding, the potential differences on the secondary winding are separated by flanges. This avoids voltage punctures in the secondary winding. In automobile manufacturing ignition coils are needed to create the required ignition temperature in the ignition plug of Otto combustion engines. Due to the section bobbin's high insulation requirements, materials with low water absorption are usually used. Possible synthetic materials are liquid crystal polymers (LCP) or polyphenyle sulphide (PPS). Section bobbins usually include pin-like or flat contacts, which are electrically connected using arc welding, soldering or resistance welding. Often a twisted wire for the coil's beginning or end is dispensed with. In some cases the design used for the individual sections deliberately reduces the number of turns on the high voltage side. A decrease in winding distribution on the high voltage side significantly reduces the risk of electrical punctures.

Pencil bobbins

Pencil bobbins (Figure 1.22) were developed only for application in automobile manufacturing. They are used as a substitute for typical section coils. The basic idea for designing pencil bobbins is to avoid high-voltage cables in the engine compartment as they are used for the traditional section coil technology. By avoiding ignition cables the puncture properties, for example in wet weather, are reduced greatly. Section coil ignition transformers are substituted by slim pencil coils, which are mounted directly above the ignition plug. The requirements for pencil coils were to enable higher temperature resistance of the ignition coil because of the proximity to the combustion engine, and to maintain the dielectric strength within the secondary part. Additionally, a slim structure of the ignition plug shaft must be ensured. Today's generation of pencil bobbins usually no longer contains sections to keep windings separated from one another. As opposed to section coils, the secondary part of the pencil coil is often inserted into the inner diameter of the primary part. After inserting all of the necessary parts into the ignition coil, the sub-assembly is often casted or overmoulded.

Cross-wound bobbins

The term cross-wound bobbin originates in the thread industry and describes a thread coil, which has a homogenous appearance by winding thread in a crosswise fashion (Figure 1.22). In electrical engineering, the term cross-wound coil often stands for two coil variants, which are very different in their actual bobbin and winding design. Cross-wound coils made from high-frequency stranded wire are often used in radio technology applications in medium wave areas or for gyro compasses. The cross-wound coil's bobbin for medium wave areas and a low winding height often only has one winding ground. By winding the wire in a crosswise fashion the winding structure stabilises itself. Bobbin flanges on the sides are only needed for high winding structures so that the turn cannot slide off axially. This type of cross coil is often used in combination with stranded wire for high operating frequencies of up to 50 MHz. The use of stranded wire reduces the collection of charges at the wire surface according to the skin effect. By winding the bobbins with very high pitches the coils become suitable for high frequencies. Bobbins for cross-wound coils are usually designed very simply.

Stator segment **Full lamination stack** **Single tooth**

Figure 1.23 Examples of a stator segment, a full lamination stack and a single tooth

Stator segment as bobbins

This type of bobbin is mainly used in the manufacturing of electric motors. The insulation is carried out by individual plastic caps that are mounted onto the segmented stator (Figure 1.23). A distinction must be made between T-segments and stator segments, which are joined in an end ring. The segmented stator is used increasingly often in modern motor manufacturing because it enables a much higher fill factor of copper in comparison to full lamination packages. Due to the easily accessible winding space, a cuneiform winding adjusted to the stator diameter can be created. Another type of application for end caps are stators with plugged single pole coils. While for stator segments, the coil is created on the lamination stack, for sheet-less single teeth the pocket insulation paper and both of the end caps are fixated in the winding machine with a tool (Figure 1.24).

Figure 1.24 Insulation concepts for stator segments (insulated with paper, endcap insulation and overmoulded, from left to right)

For this, it is almost always necessary to use a collapsing tool, because otherwise the pressure resulting from the tightly wound winding would hinder the removal of the single tooth coil. The advantage of sheet-less single teeth is that they can be used without a tooth base when combined with full lamination stacks, with only a slight decrease of copper fill factors compared to stator segments. This type of end caps is only slightly different from those wound directly wound on the lamination stack. Stator segments can be designed in one of two ways. Either the end caps are designed with long grooves according to the stack length, or the end caps are used in combination with groove insulation paper. Respectively stator segments can be overmolded as well. Possible materials are polyamide (PA), PPS or LCP synthetics. Typically, end caps offer different possibilities for contacting. The design can include grooves for the insulation displacement technology or welding hooks for resistance welding.

Full stator lamination stack

This type of bobbin is mainly used in the manufacturing of electric motors (Figure 1.23). Full insulations or synthetic end disks are usually used. These end disks are equally suitable for distributed windings (Section 1.3.5) and concentrated windings. When compared to single teeth windings, the main advantage of full lamination stacks is that the connecting wires have fewer contact points depending on the wiring topology because they are wound continuously. Just like single tooth caps, the synthetic end disks can take on various tasks. End disks are also always designed in one of two ways. Either the end disks are designed with long grooves according to the stack length and depending on the possible moulding depth and wall thickness, or the end disks can be used in combination with groove insulation paper. In principle, for groove-less end disk design one must make sure that the insulation paper used is inserted to overlap towards the disk, in order to ensure the maximal creepage distance. As with the single teeth, end disks with moulded grooves reduce the number of individual parts used. This is shown in Figure 1.25.

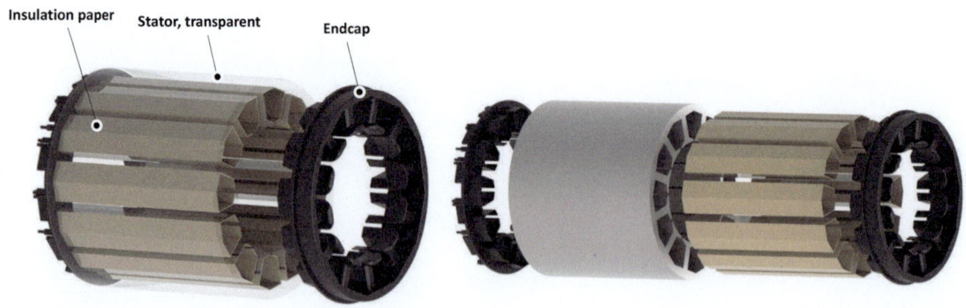

Figure 1.25 Insulation concept: endcaps/end disks with insulation paper/foil

Similar to single teeth, the grooves in the end disks can be designed to insulate the full tooth length. In this case, however, a mixture of end disk with groove and end disk without groove is created. The advantage is the elimination of a gap, which exists when two end disks with groove are used within one stator groove. It is only partly possible to create wall thicknesses comparable to insulation paper. PA, PPS, or even LCP synthetics can only partly be moulded deep enough depending on their solids content. From practical experience wall thicknesses of 0.5 mm can be achieved with a low moulding depth. This is why additional groove insulation paper is often used, especially when stators are operated with up to 48 V or even up to 800 V. In addition, insulation paper, such as Nomex, in thicknesses of 0.25 mm is common and does not limit the winding space because of its small dimensions.

As for traditional bobbins, end caps are designed to provide various contacting opportunities and functionalities, which are helpful to winding or automation (Figure 1.26). Hence, it is possible to include grooves for the insulation displacement technology or welding hooks for resistance welding of individual phases as well as to include the star point in the moulding. Another characteristic of end caps is being able to place the connecting wires of coil parts of a phase around the outer diameter of the disk. This is especially beneficial for series connection. The connecting wires of all single coils belonging to one phase can be laid without additional connecting components. If the application operates with higher voltages it is possible to separate the connecting wires of one phase's coil from the connecting wires of other phases using a groove at the outer diameter of the end disk. A high dielectric strength can be achieved using this method.

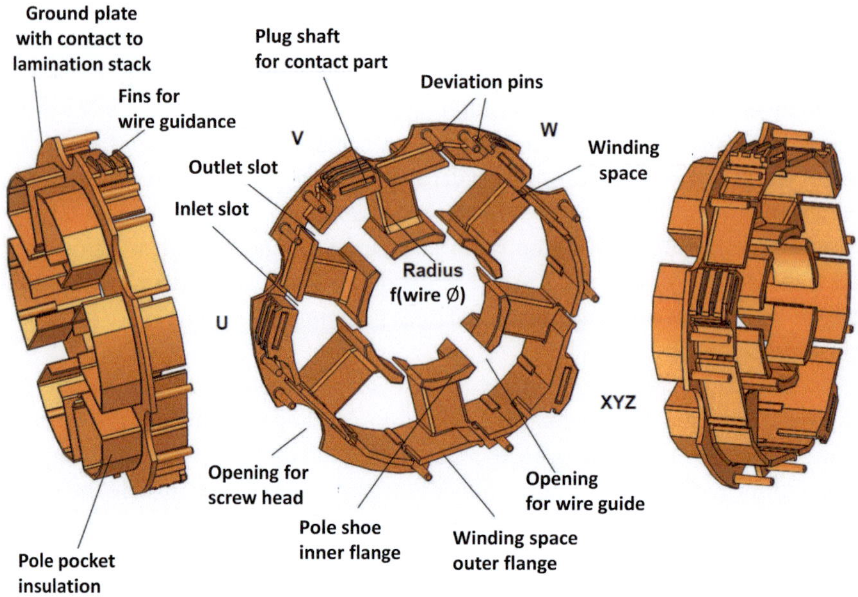

Figure 1.26 Technical features of endcaps/end disks (Picture: Technoexpert Dresden)

Single tooth bobbins

Single tooth bobbins are often used for manufacturing torque motors, in which the coils can be pushed onto lamination stacks or segmented single teeth after the winding process (Figure 1.23). Basically, single tooth bobbins are derived from simple magnet bobbins with a rectangular shape. The relatively large wall thicknesses created by the moulding process are generally used for stators with a decreased copper fill factor. The individual coils are wound as independent coils and then inserted into the stator after manufacturing. The bobbin's walls need to withstand the typical winding pressure so that the bobbin removal from the winding tool is still possible. Single tooth bobbins provide various contacting options because the operational area is often not spatially limited. Contacting methods like soldering, resistance welding and insulation displacement technology are used. Assembling onto the single tooth may be challenging due to production related deviations, considering that single tooth bobbins have a pre-defined inner contour with certain dimensions. A tight fit on the lamination stack cannot be guaranteed. Depending on the chosen material this effect can be balanced by including so-called auxiliary, respectively sacrificial bridge in the mould. These bridges, often only 0.2 mm thick bars, are moulded onto one of the slim inner sides of the bobbin in the direction of the stack's height. While mounting they are bent over or sheared off due to the oversized lamination stack. This means that a tight fit is always achieved, regardless of the lamination stack height resulting from the packeting tolerance.

Bobbins have a significant influence on the reachable degree of automation and the resulting process capability of the developed winding process. A poor bobbin design will almost inevitably lead to insufficient winding results. Efforts to increase the functionalities of the bobbin, like integrating bearing functionalities, valve interiors or wire contacts for termination, are an increasing trend. Thus, improving injection moulding technology enables the manufacturing of complex winding sub-assemblies.

1.2.5 Introduction to bobbin materials

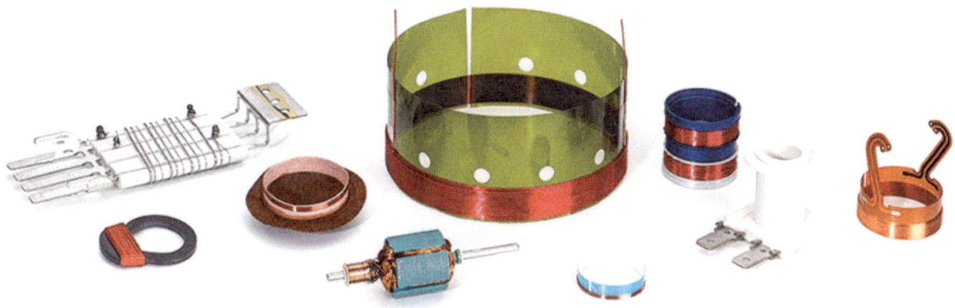

Figure 1.27 Variety of coil bobbin materials

Since the number of synthetic materials used for bobbins is very high, this section will only cover a selection. For motor coils a layered structure of metal lamination stacks is usually used as the bobbin. Compared to synthetic bobbins, these provide fewer restrictions for winding, but are limited in the variety of material. As described in the previous section, insulating the segments or full lamination stacks with plastic is common, which is why this may have an influence on motor winding process. For this reason, the focus of this section is on synthetic materials for bobbins.

Synthetic materials and their processing capabilities are continuously improved so that new possibilities for winding technology emerge. As an example, carbon fiber reinforced materials, which are used in automobile racing, provide enormous strengths. In automotive electrics and electronics, synthetic bobbins played only a minor role in the past, but are very important today. Consequently, the functional properties such as temperature, mechanical resistance and chemical stability of synthetic materials used must be well considered. When using high performance synthetics, the costs for granulate have only small impact on the overall costs. The relation of strength and density of synthetics is particularly favorable, since synthetics are superior to steel or cardboard when compared by weight. The bobbin's strength is not only relevant as winding carrier because of the winding pressure. Itis also significantly important for the wire guidance of end caps on full lamination stacks, for example. In very few cases, the use of synthetic bobbins is dispensed with for systemic reasons. These exceptions include, antennas or sensors wound onto ferrites, or metal winding carriers that are enamelled to insulate from the wire. These special types are not covered for now. In the recent past, mouldmany of the different properties required for bobbin geometries have been achieved with injection moulding technology using granulates. However, it is still very important to design bobbins suitably for synthetic materials.

When considering the synthetic strength it is often misunderstood that it is actually specific strength, meaning strength in relation to density. Put simply: when compared by weight, synthetic materials are superior to steel or cardboard! This is also why more and more functional geometries, such as moulded bearing seats for stators or anchor guides in valve bobbins are substituted by synthetic material. When choosing one of many different synthetics, the known groups are considered:

- thermosetting plastic
- elastomers
- thermoplastics.

When choosing materials suitable for bobbins, thermosetting plastics and elastomers are usually no longer considered due to inappropriate mechanical and processing properties. Standard thermoplastics are quantitatively the biggest group. The number of different materials is relatively small with polyethylenes (PE), polypropylenes (PP), polystyrene (PS) and polyvinyl chloride (PVC). Polyethylene terephthalate (PET) is often added to the group of standard synthetics due to its production volume, however according to its properties it ought to belong to the next group of technical synthetics. The share of standard synthetics for bobbin manufacturing is very small. Only synthetics like PE and PP are used in subordinated applications. Technical synthetics cover a much larger number of materials. Examples are thermoplastics like polyamide (PA), polyoxymethylene (POM) or polycarbonate (PC). A comparison of various synthetic materials is given in Table 1.4.

Table 1.4 Typical characteristics of different plastic materials according to [HUG 15]

	Density	Tensile strength	Elongation at break	Young`s Modulus	Dielectric strength	Humidity absorption	Temperature range	Thermal resistance
Unit	g/cm³	N/mm²	%	kN/mm²	kV/mm	%	°C	°C
Material	ISO 1183-1	ISO 527	ISO 527	ISO 178	IEC 60243	Norm climate		ISO 75-1/2
PE	0.95	24	>200	1	80	0	-100 … 80	48
PP	0.91	33	800	1,3	55-90	0,1	+10 … +100	65
PS	1.04	55	40	3,2	43	<0.1	-10 … 80	86
PA	1.14	50 … 80	50 …160	1.5 … 3	80 …150	2.5 …3	-40 … +100	95
PC	1.2	>60	>80	2.3	35	0.2	-40 … +135	138
POM	1.41	70	75	3	35	0.17	-20 …90	124
PVC	1.36	65	20	3	20	<0.1	-10 … +55	61
PET	1.37	74	>20	3	>70	0.2	-20 … +100	74
PEEK	1.32	160	50	3.6	20	0.2	+250	182
PSU	1.24	72	>50	2.5	>40	0.25	-40 … 150	175
PES	1.37	85	20 … 40	2.9	63	0.8	+180	215
PAI	1.45	120	5	5.8	24	2.5	+250	280
PPS	1.35	90	8	35 ... 42	50µm - 10kV	–	-20 ... 220	240
LCP	1.4	180 ... 240	3.4	10 ... 40	32	0.3	250	300

High performance synthetics are often used as a consequence of the requirements. These cover significantly higher temperature ranges exceeding 150 °C. Examples are polysulphone (PSU), polytheretherketone (PEEK), polyether sulphone (PES) or polyamide-imide (PAI). As previously mentioned, high performance synthetics have a very small market share due to their cost factor. They are characterised by one or multiple properties that are especially beneficial to the manufacturing of bobbins. Especially when using high performance synthetics for bobbins, these can be designed for high temperature applications. Ultimately, the following criteria are relevant to the choice of synthetic material for the manufacturing of bobbins:

- mechanical strength
- electrical insulation (under consideration of water absorption)
- chemical resistance
- thermal resistance
- flow properties during injection moulding
- compounding/temperature conductance capabilities.

Mechanical strength

As a main function, bobbins are used to carry the winding and to shape the coil itself. While inserting the winding in the winding space, forces resulting from the necessary wire tension can deform the bobbin. These deformations affect both the winding ground as well as the coil flanges axially. Deformations caused by the winding can influence the positioning accuracy of every turn. In addition, deformed coils usually cannot be assembled due to the limitations of the final product's installation space. Since bobbins are often only needed during coil manufacturing, and do not fulfil any further function in the final product, attempts are made to keep the wall thicknesses as thin as possible to minimise the installation space for the coil.

A ratio of a high amount of copper to a small amount of bobbin material is desirable, especially in the course of increasing demands for energy efficiency and power density. Decreasing the bobbin wall thicknesses results in the increase of the electrical fill factor of a sub-assembly, which on the other hand obviously opposes the bobbin's mechanical stability. The design of wall thicknesses always represents a weighing of interests between the bobbin material used and the available installation space, or electrical fill factor.

The characteristic value for mechanical strengths of thermoplastics or elastomers in coil winding technology is the Shore hardness, named after Albert F. Shore. The standard procedures lettered A, C, D are used and described in DIN EN ISO 868. During testing, a cone (procedure D) or frustum (procedure A, C) are pressed into the synthetic material with a certain force. The resulting depth of penetration gives the Shore value. Further characteristic mechanical quantities are ball indentation hardness, which can be defined according to DIN EN ISO 2039-1, and notched impact strength. A testing procedure for notched impact strength is defined in DIN EN ISO 179. The stiffness of synthetic materials in the elastic region can be described using Young's modulus and in the plastic region using the respective stress-strain diagram (Table 1.4). The tensile strength gives a limit for the mechanical tensile stress. Table 1.5 shows various common synthetic materials and their degrees of hardness.

By way of comparison, a car tire has a Shore hardness D of 15. Especially for the thermoplastics covered here the Shore hardness is determined by procedure D. When determining the Shore hardness temperature plays a crucial role; measurements must be taken at a temperature of 23 °C. Unfortunately, from a winding point of view it is not possible to use a harder material to meet the requirements of thin wall thicknesses. Similar to metals, the suitability for winding is given by a balanced ratio of hardness and toughness.

Table 1.5 Comparison of different plastic harnesses' according to [VDI-16]

Plastic\hardness	Shore hardness A	Shore hardness D
PE	95 ... >	40 ... 70
PP	>	65 ... 75
PS	>	80
PA 66	>	80
PC	>	82 ... 85
POM	>	79 ... 82
PPS	>	77
PVC-U	>	75 ...80
PET	>	83
PEEK	>	85
PSU	>	85

Hard synthetics, such as PPS, tend to have lower notched impact strengths. This property frequently leads to cracks during the winding process or immediately after removing the coil from the winding tool. One way of increasing the notched impact strength in synthetics is to include filler materials such as glass fibres.

Electrical insulation

Electrical insulation is often required in a context of high operating voltages for coil subassemblies. Besides the material composition of the synthetic, the influence of particles diffusing into the synthetic is relevant for the resulting insulation strength. A weakening of the insulation strength of synthetic material is caused by the tendency to absorb water. In addition, the use of filler materials can weaken the insulation strength as well.

Insulation strength is defined in ICE 60112 and is also referred to as tracking resistance for bobbins. It describes the insulation strength of the surface (creepage distance) under the influence of humidity and contamination. It defines the maximum creeping current, which may result under standardised testing conditions in a defined testing arrangement. A high tracking resistance means that barely measurable currents occur on the bobbin surface when applying corresponding high voltages. The dielectric of a creepage distance is often lower than of an equivalent length through air, even though the insulation material insulates well. The tracking resistance of an insulation material is influenced by its capability to absorb water and its behavior under thermal and ionizing stresses. Tracking resistance is determined by a comparative tracking index (CTI) value and is measured by a proof tracking index (PTI) value. Using an apparatus as defined by IEC 60857 and DIN EN 60112, a liquid is applied to the bobbin surface while a voltage is applied to the bobbin. The CTI value describes the voltage limit for no conductivity in the creepage distance. The corresponding failure criterion is a creeping current of more than 0.5 A. Besides tracking resistance often the material's insulation class is provided. It is defined by IEC 85.

Chemical resistance

Especially in automobile manufacturing, bobbins come in contact with substances, which might negatively influence the strength, insulation strength, or temperature resistance. A bobbin is chemically resistant, if its characteristic mechanical, physical and chemical properties remain constant, even if it comes in contact with a chemical substance for a long period of time. Since this ideal situation practically never occurs, a material is technically "resistant", if it is only affected very slowly. A synthetic bobbin is chemically resistant, if it keeps its characteristic properties for a defined and suitable amount of time, or within certain operating conditions. The chemical resistance plays a significant role, especially, when the coil system is casted, impregnated, or overmolded. Beforehand, the compatibility of the different materials has to be considered. Common applications are so-called wet runners. Surrounding media, such as petrols, oils or even de-icing salt can lead to extreme damage of the synthetic.

Thermal resistance

Electromechanical sub-assemblies are not only warmed up by ambient temperature. Due to the increasingly refined designs, coils frequently have operating temperatures of around 150 °C at normal ambient temperatures. To ensure bobbins do not deform, melt or even burn at high temperatures, synthetic materials such as PEEK or PPS were developed. PA, for example, only has a thermal resistance of 130 °C, which would not be sufficient for applications in mechatronic systems. By utilizing a PPS, mechatronic systems can be operated at temperatures of up to 200 °C. The permanent thermal resistance is labelled with the letter A, B, C etc. These letters stand for the highest permanent temperature for insulation materials defined in IEC 60034-1 (Table 1.6).

Table 1.6 Insulation classes for electric insulation materials according to IEC 60034-1

Insulation class	Max. permanent application temperature in °C
Y	90
A	105
E	120
B	130
F	155
H	180

Manufacturability/flow properties of injection moulding

As previously mentioned, the synthetic material used must be adjusted to the coil's operation location and type. This mainly influences the construction of the injection moulding tool. However, while PA materials can be manufactured with low temperature injection moulding tools, when using PPS, operation temperatures of the injection moulding tool as well as the injection channel have to be pre-heated accordingly to the processing temperature of around 150 °C. The tool temperatures may reach values of around 240 °C when using PEEK, which directly affects the tool costs. Adding glass fibers or mica has a negative influence on the flow behavior. Using software, a flow analysis of the injection moulding material is often performed in advance to avoid the risk of a tool misconstructionmould. With the help of computerised analysis, the material flow behavior in the tool can be determined and a possibly early solidification can be established, which would lead to bad weld line strengths. Polyamides are best suited for the processing with injection moulding. The different viscosities of various synthetic granulates results in unequal requirements for the tool separation point and the use of sliders. If the thaw has a low viscosity, an unclear separation of the tools may cause feathering. The resulting burr can be the cause of partial discharges, especially in the manufacturing of high voltage bobbins, wire damages respectively winding errors, which lead to premature malfunction. For an accurate winding process often very small geometries are needed to place the wire in a certain position. It is not uncommon, that radii in sizes R = 0.1 mm must be formed. This also limits the possible material selection. The surface resulting from the injection moulding process is also relevant for the winding of pencil coils, for example. When winding at a pitch, the surface must provide a certain roughness so that the turn does not slip.

Compounding/temperature conductance capabilities

Compounding capability describes a granulate's ability to mix with foreign substances. Compounding is a term from synthetic technology, which is equal to processing and refining the synthetic material by mixing in additives (filler materials, additive glass fibers, etc.) to specifically optimise certain properties. Better temperature conductance capabilities can be achieved by mixing in quartzes or mica. To compound the material, it is mixed and ground or liquefied by an extruder. The variety of possible additives is large and reaches from low-melting glass or metal alloys to heat conducting materials. However, these additives impair the flow behavior of the material. For synthetic materials the coil properties during operation and the manufacturing properties must be weighed against each other. These negative effects can be reduced by prior grinding of the material and subsequently compounding. This is utilised for glass fibers and reduces the viscosity significantly. Based on development of mechatronic systems with higher power densities, compounds with higher temperature conductance capabilities have gained significance. For servo motors power densities of 1.5 kW/dm3 are common. For higher power densities external cooling by air or even water is often required. To reduce these cooling efforts, one approaches targets to improve the temperature transfer from the winding to the housing. This can be achieved by mixing temperature conductors into the synthetic granulate for the winding carrier. As mentioned before, compounds change multiple physical properties of winding carrier. While adding a small amount of filler material increases the strength at first, it starts to decrease again with amounts of over 50 %. Generally, no direct recommendation of a certain material can be given. The choice of a suitable synthetic and optional filler material always depends on the application and the part's geometry.

Friction properties

Friction properties of synthetic materials play a key role in winding processes because they directly influence the wire positioning on the bobbin. For characterisation purposes, friction coefficients are given, which can be determined by different tribological experimental setups. Besides the chosen material, bobbin friction properties depend on the surface roughness, the contact pressure of the friction partners and the temperature. Pencil coils are roughened so that the wires cannot slip. An exception for pencil coils is overmolding the winding ground with a softer layer. This is often done by using a two-component injection mould to apply a mechanically flexible coating. A collection of friction coefficients of different synthetics is shown in Table 1.7. These friction coefficients were determined with a contact pressure of 0.1 Mpa and a speed of 0.15 m/s with a ring-ring tribometer.

Table 1.7 Friction coefficients for different plastic combinations [BASF-14]

	PA 6	PA 6- GF	PA 66 mod.	PA 66-GF	PBT	POM
PA 6	0.65 … 0.85					
PA 6- GF	0.45 … 0.55	0.35 … 0.45				
PA 66 mod.	0.5 … 0.7	0.35 … 0.45	0.15 .. 0.2			
PA 66-GF	0.55 … 0.75	0.46 … 0.6	0.2 … 0.26	0.32 … 0.44		
PBT	0.7 … 0.9	0.65 … 0.85	0.42 .. 0.57	0.6 ... 0.8	0.6 … 0.8	
POM	0.5 … 0.7	0.35 … 0.5	0.2 … 0.3	0.22 … 0.3	0.4 .. 0.6	0.3 .. 0.5

The continuous development of material systems, primarily of synthetic materials, enables the use of coils with stronger mechanical, thermal and chemical stresses. The permissible manufacturing tolerances for coil production are increasingly limited and require efficient manufacturing processes and product designs. The properties relevant to the winding process are rigidity and dimensional precision of bobbins, also in terms of winding flanges for smaller wall thicknesses as well as the friction properties of the surface. Due to the improved possibilities of synthetic processing, guiding geometries, such as grooves, can be included in the winding ground to support the specific guidance of the wire in the groove.

1.2.6 Coil design influences on manufacturing properties

General aspects

The purpose of this section is to illustrate the interaction between coil design and its manufacturing properties. In addition to the influencing factors of the bobbin design, the material influences are described in their impacts on winding technology. After an introduction of general aspects, the specific properties of certain coil types are covered.

Primarily, the desired final shape of the wire is crucial for the winding capabilities of bobbins. For complex winding schemes, the first layer of the winding is therefore supported by a structure in the winding ground, in the shape of grooves, in order to guarantee a correct positioning from the beginning of the winding process. For cylinder coils, disk coils, or ring core coils, accessibility of the winding space is given. For motor windings with lamination stacks this is not the case. To enable sufficient accessibility of the winding space, aside from the electromagnetic design, the desired wire cross section has to fit through the groove between individual teeth. Depending on the wire gauge and its material, different minimal bending radii result for each wire, which have to be considered when guiding the wire into the winding space via start-wire and end-wire pockets or guiding around a non-circular bobbin. As a general rule of application, the minimal bending radius roughly equals the wire gauge. Flanges are often used to limit the winding space, however, these have to be capable of withstanding the axial pressure of the winding. As an approximation, the angle between grooves, or the winding ground and the winding flanges should be between 60° and 90°. There is evidence that this range is favourable for direct winding with individual wires (Figure 1.28).

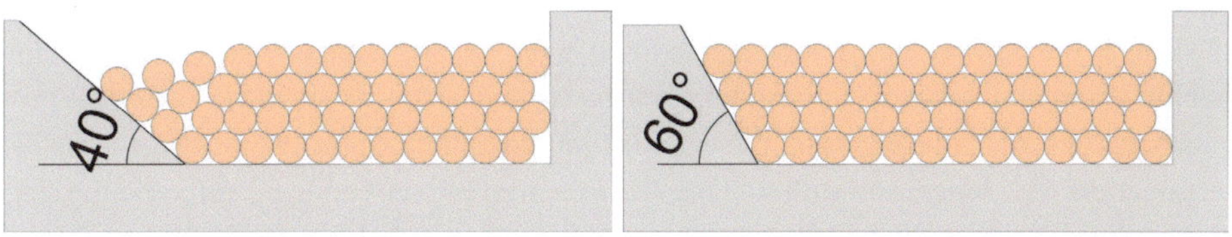

Figure 1.28 Comparison of an unsuitable (40°) and a suitable (60°) angle for the inner groove geometry

An illustration of influencing factors for a successful winding result is given in the shape of an Ishikawa diagram in Figure 1.29. The material and the winding method including geometric aspects are important factors; but further properties, including those of the human (operator), the machine, the environment (climate) and measurements have to be taken into account as well.

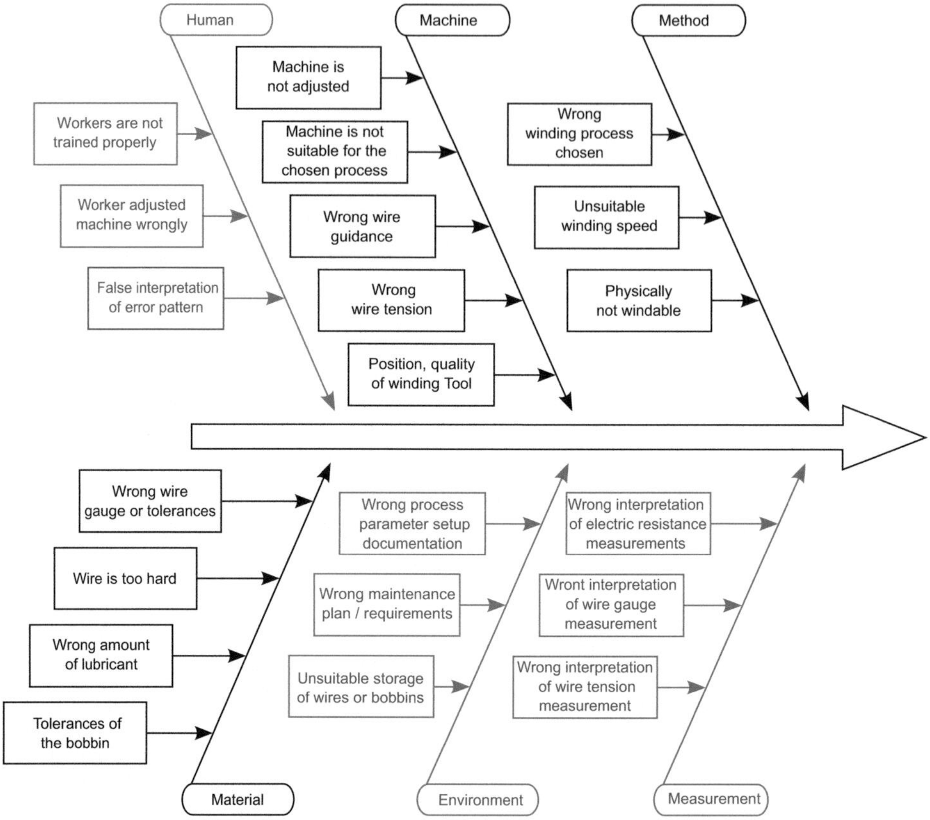

Figure 1.29 Influencing factors on winding process and good

Due to the number of influencing factors it is important to choose the coil design and bobbin material suitable for the winding process, as well as for the up and downstream processes. Aside from winding itself, the wire fixation possibilities prior to winding, as well as termination, wire separation without wire remains from the start and end wire, and the contacting as a downstream process have to be considered for the bobbin design. Depending on the contacting process, it may be necessary to wrap the wire to a solder tail or fixating hooks, or forked flags for resistance welding. The same applies for insulation displacement grooves or contacting of a lamination stack stator via a lead frame for the start and end wire or the possibility to place connecting wires of individual phases/coils in the end disk.

The task of inlet clamps for the wire, also referred to as start wire pocket, is to distance the winding package from the start wire. The first turn of the winding must always be guided to the winding ground to establish a winding structure for procedural reasons. The last turn is usually located on the winding's surface. A missing spatial separation of the start wire running from the front side to the winding ground and to the winding package leads to physical contact between the start and end wire. This may lead to a premature malfunction of the coil due to the high potential difference at this point.

Placed over the winding process, the use of characteristic features gives additional possibilities of automated quality control, be it according to Poka Yoke or for the design of the bobbin (Section 4.4.1). Typically, the highest risk potential of a bobbin for the winding process is in its manufacturing precision. A burr on the bobbin caused by the manufacturing process may interfere with the winding process. Shape deviations of the bobbin are also critical, regardless of whether they were caused by the construction, the manufacturing process or the used tool. Alternatively, environmental influences can cause such deviations, when, for example, synthetic materials swell because of the ambient humidity at the production location. Apart from the shape, the bobbin's rigidity is also crucial, since the constriction effect of the applied winding can lead to a deformation of the core or flanges, or may break the bobbin during the winding process. Countermeasures to rule out the malfunction of semi-finished products are the integration of preliminary examinations, such as the winding readiness check in terms of the bobbin's dimensional precision, storing the semi-finished products under standardised conditions, or a mould flow analysis of the bobbin construction and manufacturing properties. Typical error patterns for the winding process are incorrect winding geometries, such as incorrect applications of the winding scheme, incorrect electrical properties, such as resistance or inductance, damaged wire or wire insulation, damaged bobbins, false bobbin/winding dimensions, such as outer and inner dimensions of the coil, and, in extreme cases, a broken bobbin or a torn wire.

The economic significance of a stable process design is illustrated in Figure 1.30. According to this figure, for each progress in the development phases of a product the costs for error correction are multiplied by a factor of ten. Consequently, an inclusion of manufacturing aspects during in the development phase can reduce the error costs by a factor of one hundred.

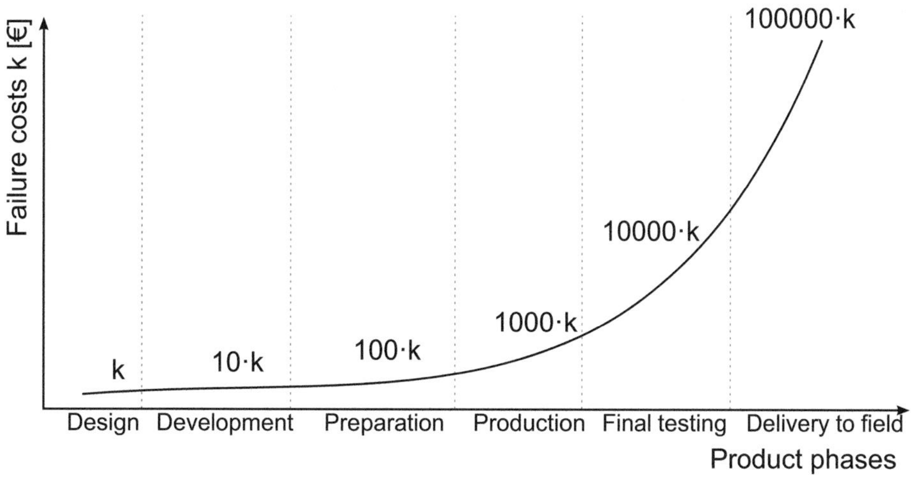

Figure 1.30 Stages of product development and related costs of error correction

In the following sections the conditions and requirements of bobbins for winding technology are covered in more detail, respective winding processes will be introduced in Chapter 3.

R_M bobbins or pot core bobbins

In terms of winding technology, R_M bobbins with axial contacting pins are difficult to process, because a pivoting wire guide nozzle has to be used (Figure 1.21). R_M coils are manufactured using the bifilar wire technology because of the high frequency region (Section 3.4.4). Wires from two wire guide nozzles are processed simultaneously. For this, the distance between the pins of the bobbin has to be matched with the distance between the nozzles of the winding machine. Contacting the wire with contacting pins is almost always done by soldering.

Relay bobbins

From the perspective of winding technology, relay bobbins are easy to wind. Usually, the contacts located radially are easily wrapped. Bobbins without continuous contacting pins at the coil flange have the purpose to release tension from the wire. These wires are often wrapped and have gauges of less than 0.08 mm in combination with twisted start and end wires. Twisting the wire creates a twisted bundle of three to five times the wire cross section (Figures 2.9/2.10) and reduces the risk of a start wire rupture. For medium voltages, relay coils are made from wires with 0.03 to 0.2 mm. Relay coils for medium voltages can be wound with 20.000 turns when using very thin wires. Relay bobbins tend to be small due to the limited operational space. Core diameters of less than 2 mm are not uncommon. Because of their small size, these bobbins can be wound with high speeds of up to 25.000 revolutions per minute. Relay bobbins used in combination with medium or high voltage application of 400 V are often designed to feature a section for the incoming start wire. In the automotive industry, typical numbers of turns are up to 1000 turns.

Valve bobbins

In terms of winding technology valve bobbins are unproblematic. In the automotive industry the bobbins are often designed for orthocyclical windings because of the limited spatial conditions (Section 3.1.2). Since the bobbin flanges and the winding ground usually have thin wall thicknesses, one is frequently forced to work with collapsing winding tools and counter-bearings for the accommodation of the bobbin. As valve bobbins are often exposed to high temperatures, especially in automotive manufacturing, they are made from temperature resistant materials, such as PPS or PEEK. The contacting processes are limited to temperature resistant types of connections, such as arc welding, resistance welding or insulation displacement technology.

Section bobbins

From the perspective of winding technology, the bobbin's section transition is very challenging. Each section has a transition area to the neighbouring section in order to continue the winding in the following section. In this transition area, the winding can only be continued at much slower winding speeds. To achieve this transition an interruption of the actual winding process is necessary. The winding is very time-consuming, since for ignition coils up to 30.000 turns have to be applied. Therefore, the aim is to wind the coils at the highest possible speed. But due to the non-circular bobbin shape the process speed is restricted. The causes are different wire speeds, which change constantly within one turn. This results in a partial loose wire every turn, which in the worst case, can create a wire loop within the winding. These loops can create a high voltage potential within a section and may cause a premature malfunction of the ignition coil. Low quality of the moulding parts, for example in the form of burrs, is a key element in the quality required. Since wire gauges from around 30 µm to around 100 µm are used, the risk of damaging the wire or the wire getting stuck on the moulding burr, especially at the section transition, is very high.

Pencil bobbins

The shape of pencil bobbins implicates that fundamental changes have to be made to the wire traversing. The winding is produced using the so called "pilger step process" or "pitched winding process" (Section 3.1.2). This means that winding sections with a length of around 2 mm are applied at a maximum angle of 12°. The wire guide moves in the direction of the traverse with certain steps cyclically the so-called pilger steps. The most common windings consist of multiple partial windings based on the pilger step process. Pencil coils meet the requirement of reducing the potential by applying the winding step-by-step, without the necessity of using separate winding section. To create such a pitched winding, first it is necessary to fill the rectangular winding space on the low-voltage side, so that a pitched winding surface is created, the so-called starting triangle. In the following cylindrical part of the winding, a cylindrical winding section is created with the pilger steps and constant traverse progress. Towards the end the traverse progress is enlarged, so that a conical winding is created on the high-voltage side. The conically sloping part towards the high-voltage side causes better insulation capabilities via smaller electrical potentials within the coil, similar to reducing the windings in a section coil. Quality requirements for pencil coils are similar to those for section coils. The bobbin is manufactured from high-grade materials, such as PEEK, PPS or LCP, due to the enormous thermal stresses. Although the winding is applied with a pitch, individual windings must not slip off. Therefore, the winding ground is manufactured with a certain roughness, so that at least the lower winding has a tight fit on the bobbin. For section bobbins, special attention should be paid to a clean separation of the injection moulding tools. Even a small moulding burr leads to wire damages and a random winding structure. Based on the previous description it can be derived that winding of pencil bobbins is very challenging. A current trend in ignition technology is the installation of both pencil and section coils as direct plug-in ignition units directly above the ignition plug in combustion engines.

Stator segment bobbins

Stator segment bobbins are often manufactured using linear winding technology (Section 3.3.4) or flyer winding technology (Section 3.3.5) for concentrated windings of electrical motors. Winding single teeth with end caps is often problematic due to geometrically caused edges and aspect ratios. An orthocyclic winding is often required and can only be created in combination with grooves in the edges of the winding space and a smaller winding speed. For single teeth, which are joined together to assemble a stator, creating wire crossings on the long side must be avoided, since this would reduce the copper fill factor in the stator groove. The end cap is usually designed such that the winding and layer steps are created on the short, meaning front-facing, side of the single tooth. This measure leads to optimal copper fill factors. Generally, the number of contact points to connect the stator is much larger compared to winding on full lamination stacks. However, this also scales with the segment size. In an ideal case, the segment equals the phase, because then the number of contact point equals those of the full lamination stacks. The winding speed, however, is limited by the non-circular winding geometry.

Full stator lamination stack

Most full stator lamination stacks are often manufactured using the insert technology (Section 3.3.8) to manufacture distributed windings or using the needle winding technology for concentrated windings. These two processes have limited copper fill factors due to the restricted spatial conditions while inserting the winding. Winding on full stator lamination stacks with full insulation is usually only possible using the needle winding technology (Section 3.3.6). An orthocyclic winding can only be created in combination with a grooved contour in the edges of the winding space and a small winding speed. For the orthocyclic winding of a full stator lamination stack, creating wire crossings on the long sides in the groove area must be avoided. The end disks, just like the single teeth, have to be designed so that steps in windings and layers are created on the short, meaning the front-facing side of the single tooth. This measure creates an optimal copper fill factor of the groove. Generally, the design of full lamination stacks using synthetic end disks has significant advantages in terms of production and costs, even if one considers the reduced copper fill factor. The number of contact points is much lower compared to stator segment coils, since with ideal accessibility the three phases of the stator can each be wound continuously. However, in comparison, the poor accessibility of the grooves limits the winding speed.

Single tooth bobbins

Single tooth coils are often manufactured using linear winding technology (Section 3.3.4) or flyer winding technology (Section 3.3.5) for concentrated windings of electrical motors. An orthocyclic winding can only be created in combination with a grooved contour in the edges of the winding space and a small winding speed. Creating wire crossings on the long sides must also be avoided in the orthocyclic winding of a single tooth bobbin. However, the number of contact points to contact the stator is significantly larger in comparison to full lamination stacks. Additionally, the non-circular winding geometry limits the technically achievable winding speed.

Cross-wound bobbins

The main difference between a cross-wound coil and a conventionally wound circular coil is in the winding process. Cross-wound coils are wound with a pitch, which is the equivalent of a multiple of the wire gauge (cross-wound winding, Section 3.1.2), so that a cross-wounded winding is created. In addition, the term cross-wound coil is linked to bobbins wound with crossing wires. The bobbin describes the shape, as for a gyro compass, but is not a cross-wound bobbin. Most of these bobbins are especially made for each application. Coils wound crosswise can be used as step motors for temperature or tank gauges in automobile manufacturing. In terms of winding technology, cross-wound bobbins for step motors are quite challenging. Because there is no opening to fixate the bobbin during the winding process, they must always be fixated with a counter-bearing.

The bobbin has many ways of exerting influence on the properties of the winding process. The parallel development of both the product and the production processes is necessary to avoid later modifications of the product design and to increase the product quality and economics of the winding process (Simultaneous engineering).

1.3 Introduction to electric motors

Figure 1.31 Wound externally-grooved brushless DC-Stator

All kinds of electric motors need electromagnetic excitation via a winding in the rotor or stator. Coil winding technology, therefore, is a key technology for electric motors. Due to the variety of electric motor designs, different challenges for winding technology occur, which is why electric motors have a dedicated sub-section. As an introduction, a definition as well as a discussion on efficiency is given. The different operating principles are explained and classified based on functional principle, operating behaviour and structure. An overview of the different application areas provides, besides information about general requirements and those specific to the application, insights into wire gauges, the number of turns and the operating voltages. At the end, key properties of concentrated and distributed windings of electric motors are introduced. This book does not focus on electromagnetic winding design, but shows the winding parameters from a manufacturing point of view.

This section is aimed at readers, who would like to understand the basics of electric motors and the challenges specific to applications in more detail. Additionally, an introduction to different variants of windings of electric motors is given.

1.3.1 Definition of electric motors

The term rotating electric machine, which is used as a synonym, originates from electrical power engineering and generally describes electric motors as energy transformers. They either transform mechanical to electrical energy (generator mode) or electrical to mechanical energy (motor mode). The main components of motors are the mechanically rotating part, the so-called rotor, and the mechanically stationary part, the so-called stator. The stator consists of a sheet metal package, which carries the excitation winding made out of electrically insulated copper coils. When manufacturing small motors, often iron-less stator windings from self-bonding enameled wire are used. The structure of the rotor varies depending on the type of motor and functional principle, which is introduced in Section 1.3.2. Generally, the motion can be described as an electromagnetic force applied on the rotor in the magnetic field of the stator. Depending on the motor type, different forces may occur. Most electric motors are based on the force delivery by the force of reluctance or Maxwell's surface force.

A practical classification of the motor technology can be made by the principle of operation and the power class. The classification comprises miniature motors (up to 1 W), small-sized motors (1 W to 0.75 kW), medium-sized motors (0.75 to 375 kW) and large-sized motors (> 375 kW). Typical applications for miniature motors are in micro technology or micro system technology and medical technology. Direct current (DC) motors are often used due to the limitation of the operation space. Small-sized motors are often used in the consumer product area, but also as small actuators in cars. For this class, DC motors are used as well, in addition to asynchronous motors. The number of miniature and small-sized motors sold in 2010 was around 300 million units, with a production value of 3.932 million Euros in the European Union (EU) [Wai-11]. Overall, this segment has the largest sales numbers. The medium power segment is used in both traditional automation technology and drive units for electromobility. Three-phased alternating current (AC) asynchronous and synchronous motors are dominant here. With sales of 30 million units in 2010 and a production value of 4.704 million Euros in the EU, this power segment has the highest sales volume and is therefore the most relevant for winding technology. Motor units with a power rating above 375 kW were sold 0.6 million times in 2010 with a production value of 1.206 million Euros in the EU.

Besides quantities and applications, developers are increasingly shifting their focus onto the efficiency of electric motors. Efficiency or losses of electric motors rely heavily on the motor type and the individual design. For high-torque applications, the losses in the excitation winding are dominant, while for high-speed applications the eddy current losses in the lamination stack are crucial. To generate a higher torque, the current in the excitation winding has to be increased. Due to the quadratic relation between the electrical losses of the winding and the current, winding losses are dominant for high torques. For higher speeds of a motor, the orbital frequency of the exciter field has to be raised, which is often realised with the inverter and without constructive changes. Higher frequencies of the magnetic field lead to more frequent remagnetisation of the electrical sheets, which leads to increased magnetization losses as well as eddy current losses. These phenomena are covered in the appropriate literature on electric motors and are treated in more detail there. Consequently, this book shall focus on the production aspects of electric motors [Fi-13].

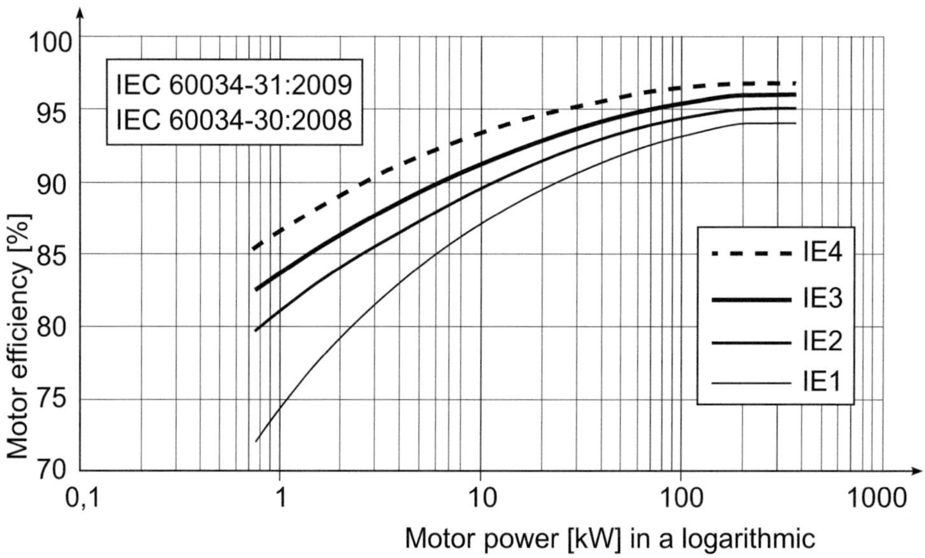

Figure 1.32 Motor efficiency for different powers according to the IEC efficiency classes

Around 40 % of the electric energy produced worldwide is used in electric motors. A small share of this, between 10 and 20 % depending on the efficiency, is transformed into heat and therefore lost. The bigger share is supplied to the respective application as mechanical energy. The medium power segment alone accounts for 68 % of the consumption of all electric motors [Wai-11, p. 30]. As a consequence, the legislation has initiated stricter standards and regulations, which is why the new energy efficiency class IE4 was introduced. The new energy efficiency class describes a minimum efficiency, which the motor has to achieve according to its power segment (Figure 1.32). By comparison, typical characteristic curves for different motor types are summarised in Figure 1.33.

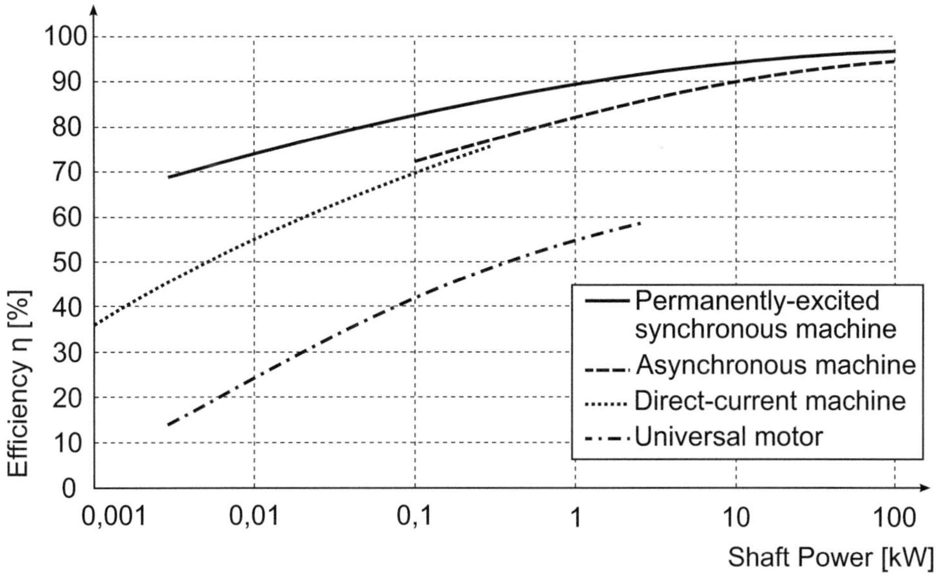

Figure 1.33 Motor efficiency for different powers and motor types

> Besides electromagnetic design, the manufacturing properties of the excitation winding of electric motors are crucial in terms of efficiency. The motor design, however, dominates the power properties.

1.3.2 Functional principles of electric motors

The purpose of this sub-section is to provide an overview of the different functional principles of electric motors and to classify the various motor types. The primary classification can be made according to the type of supply voltage. A distinction is made between DC motors, single-phase AC motors, which are also referred to as universal motor, and three-phase AC motors.

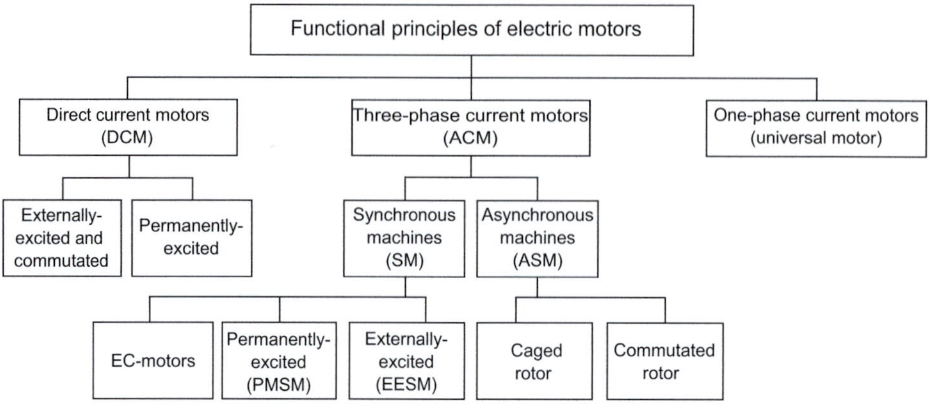

Figure 1.34 Overview of motor principles

The secondary classification is determined by the principle of operation and the last step of classification is determined by the constructive design of the motor. A basic differentiation is illustrated in Figure 1.34, as well as the constructive design in Figure 1.35.

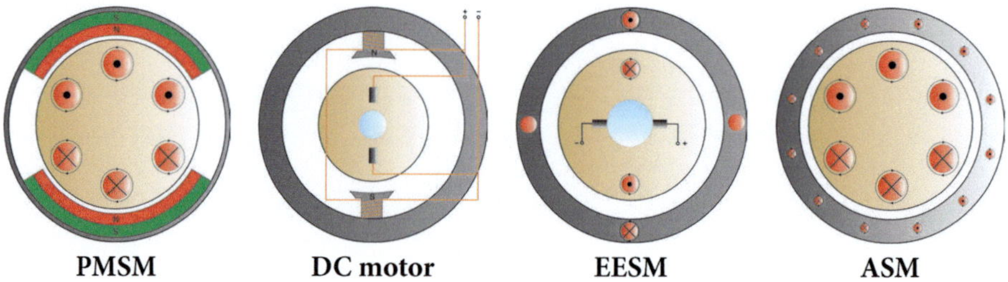

Figure 1.35 Design for different motor principles

DC motor

As described in Section 1.3.1, electromagnetic propulsion typically requires a magnetic field in the stator, and a magnetic field in the rotor. For the motor's motion, one of the magnetic fields has to move with the desired speed of the motor. The stator of a DC motor creates a constant magnetic field using permanent magnets or an energised winding. Due to this principle, a DC motor does not have to be built with a lamination stack structure. However, DC motors with higher efficiency requirements are built using lamination stacks. The needed rotation of the magnetic field is achieved by switching coils on the rotor. To utilise the Lorentz force, the rotor winding has to be energised, so that it moves perpendicular to the exciter field. The switching can be done with a commutator on the rotor, depending on the modality. The commutator contains bars, which are contacted with carbon brushes (Figure 1.35). The rotor's coils are switched to the desired polarity by the commutator. If both rotor and stator are built with windings, there are additional possibilities for interconnecting rotor and stator coils, which can be chosen depending on the desired operating properties.

In this context, a series circuit means a series connection of stator winding, also referred to as excitation winding, and rotor winding, also referred to as armature winding. An advantage of this interconnection is that this configuration can also be operated with an alternating current, where the stator and rotor field each use the fitting half wave of the alternating current. However, if the stator winding resistance is too high, this limits the available current in the rotor and stator winding. The torque characteristic of this motor shows, as illustrated in Figure 1.36, a strong dependency on the speed. A low counter induction occurs at low speeds, which is why a large current can flow through the rotor winding and therefore also the stator winding. A stronger weakening of the stator field by the rotor winding occurs at higher speeds, which is why the torque decreases. An application typical for high torque and low speed is the starter motor in passenger cars.

A shunt circuit is equivalent to a parallel connection of stator and rotor winding. In this configuration the current for the stator winding is limited due to its higher resistance and inductance. The use of an alternating current is not suitable for this configuration due to the different inductances. By limiting the current in the stator winding, the torque characteristic, as shown in Figure 1.36, is almost independent of the speed. Consequently, they are suitable for applications with constant speeds but variable loads, such as conveyor belts. The achievable torque is limited through the maximum allowed rotor current. This is why cooling measures are often applied for shunt circuit motors, such as cooling fins in the housing or a fan on the shaft.

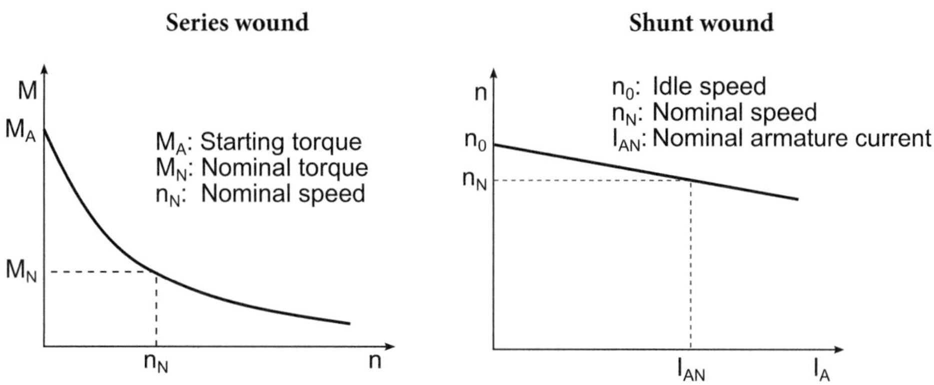

Figure 1.36 Characteristics of shunt circuit and parallel connection

A special type of DC motor for smaller power ranges up to 0,1 kW can cope with air coils for the rotor winding and therefore does not need a sheet metal package. High speeds with high efficiencies can be reached due to omission of iron losses and the lower moment inertia of the so-called bell armature motor.

The brushless DC motor (BLDC), an electrically commutated DC motor, operates, contrary to its name, according to the principle of alternating current synchronous motors, but has a control behavior, which is similar to DC motors. Due to its structure it is explained in the following sections.

Alternating current motors (universal motor)

The structure of single-phase AC motors, also referred to as universal motors, is comparable to the aforementioned structure of DC series motors. They can be differentiated by the shape of the lamination stack and the use of alternating current for the stator winding. For the universal motor, the grooves are shaped as so-called pole shoes to receive the stator winding, as shown in Figure 1.37. As in all AC motors, both rotor and stator are built out of a lamination stack to limit the occurring eddy currents. Due to this structure, the motor is suitable for both operation with alternating or direct current, which is how the name was derived. The power losses in DC motors with a series circuit would be around 15 % higher than in a comparable universal motor. The structure for higher power changes because multiple poles are used instead of one pole. Consequently, the stator is designed with a traditional lamination stack structure with multiple grooves for the stator winding. This design is used in power tools, for example. Special types of universal motors are the capacitor motor and the shaded-pole motor, which basically work according to the asynchronous motor's principle. This principle is characterised by a phase shift of the rotor's and stator's magnetic field. The capacitor motor uses a capacitor between the windings of stator and rotor, so that a phase-shift is created. Typically, capacitor motors are used in household appliances, like drives for refrigerator compressors, roller shutters or marquees. The shaded-pole motor is switched in a way that each pole is split with the use of a shading coil. One half of the pole is equipped with a shading ring, causing a delay in the creation of the magnetic field, and a phase-shift occurs. Typical applications for shaded-pole motors are pump drives for washing machines or dishwashers with a power of up to 100 W [Rud-08, p. 229].

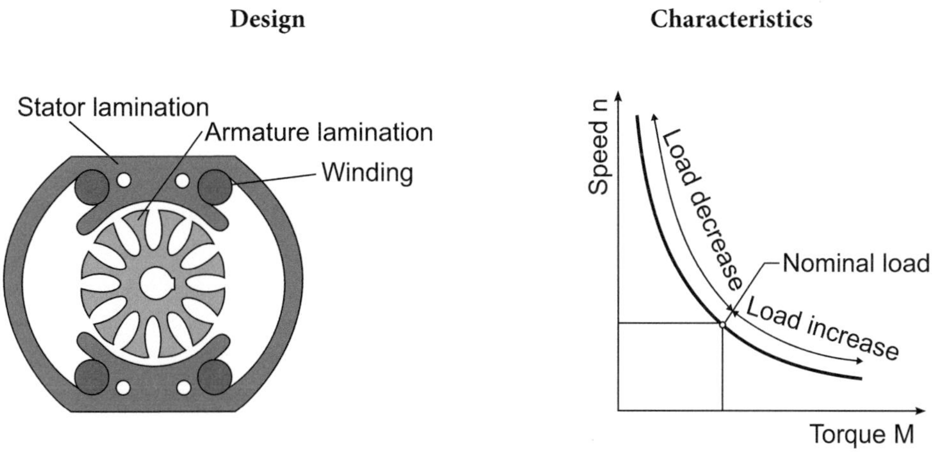

Figure 1.37 Design and characteristics of a universal motor

Three-phase AC synchronous motors

As opposed to single-phase AC motors, multiple phases of alternating current are needed for the synchronous principle. As the name suggests, the magnetic fields of rotor and stator are in phase. The secondary classification determines the way the rotor field is created. Either permanent magnets can be installed on the rotor for a permanently excited synchronous motor, or a winding is applied, which is then energised using a collector ring for an externally excited synchronous motor. A converter determines the frequency of the stator field to control the drive. The current defines the torque applied to the rotor. Due to these different effects, the control is highly dynamic even for higher power segments. Compared to DC motors, a much lower excitation power needs to be transferred via the collector rings or can be neglected completely for permanently excited motors. This results in higher efficiencies of up to 98 %. Due to the fixed coupling of the rotor and stator field phases, these motors are suitable for applications with constant speeds independent from the load, such as in actuators. Alternatively, a control using an inverter and a speed sensor can be integrated to other motor principles in order to achieve higher speed stability. Especially for applications of lower power and substantial spatial limitations, synchronous motors are used because of their compactness. However, these motors are also used for applications with high efficiency requirements in high power segments, such as ship propulsion drives (20 MW) or wind turbine generators (10 MW). A special type of permanently synchronous motor is the transverse flux motor. As the name suggests, the stator coils are arranged in the direction of the circumference. Consequently, the magnetic flux of the stator is perpendicular (transverse) to the rotational plane and therefore parallel to the axis of rotation. The advantage of this design is that the electrical and magnetic circuit of the motor are independent of one another and can therefore be designed separately. The stator winding heads, the parts of the winding that are outside the groove, which normally contribute some to the motor's losses, are omitted.

The special design enables the use of finer pole splits, in particular, which leads to higher torques even at lower speeds. This is why these motors are especially suitable for applications with low speeds but high torques, without having to use a gearbox. The reluctance machine is a special form of the externally excited synchronous motor. Reluctance is the term used to describe a magnetic resistance within a closed magnetic circuit (Section 1.2.2). The basic principle of the motor type is based on the fact that a magnetic circuit will always try to achieve the lowest possible magnetic resistance. For example, if a magnetic circuit has a loose segment and a magnetic flux occurs it will centre the position of the segment to minimise the magnetic resistance. The structure of these motors, therefore, is simple. The rotor consists of a laminated structure, which is aligned to the respective stator pole via the reluctance force. The losses therefore only occur in the stator field, which is why the reluctance motor achieves high efficiencies, but can only transfer small torques. If the rotor shape is designed in the shape of a tooth and the number of stator teeth is higher than of the number of rotor teeth, this design is often referred to as a switched reluctance motor. During operation, the single-teeth are energised in such a way as to attract the next rotor tooth. Right before the rotor tooth has reached its aligned position, the exciter field is switched to the next tooth to maintain the motion. Due to this principle, the rotor position is always known and works without additional sensor components. Due to the low torque at higher power and low efficiency at lower power, these motors are only used for medium power applications. Thanks to its robust structure they are often used in rough environmental conditions. In addition, they are used for applications in hybrid car powertrains as parallel hybrid motor. In contrast to other motor types, they do not create power losses when the combustion engine takes over due to the missing magnetic active elements.

The brushless DC motor, also referred to as electronically commutated motor, generally has permanent magnets and a stator design with coils, usually in the shape of a three-stranded AC winding. Simple variants may only have two- or single-stranded windings. The particular characteristic of the motor is that the AC winding can be controlled with a frequency converter, which provides a rectangular modulation. The motor's behaviour can be taken into account and adjusted due to the specific control of the voltage according to rotor position, speed or torque. Therefore, the rotor's position and speed have to be known. These can be determined by sensors in the rotor's magnetic field or optical systems on the rotor shaft. Alternatively, the rotor position can be captured using the induction in the stator coils at a certain minimal speed. Typical applications for this dynamic torques at low speeds are small drives for fans, compressors or servo motors.

AC asynchronous motors

Asynchronous motors consist of a stator with the aforementioned AC winding and a rotor typically built as a cage or rarely with coils, which are both grounded via the commutator. This principle generally includes a phase-shift between the magnetic field of the stator and the rotor. The phase-shift is referred to as slip and leads to the rotor always rotating slightly behind the stator field. Due to the rotation of the stator field, a current is induced in the cage of the rotor or the commutated coils. This induction leads to an electrical current, which in turn creates a magnetic field. Once the magnetic field is formed, the interaction between the magnetic field of the rotor and the current in the stator winding leads to a Lorentz force.

This results, analogue to the other motor principles, in a torque for the rotor. Due to the possibility of commutation of the rotor winding the asynchronous motor can show the already mentioned behaviour of a shunt and a series circuit. The cage rotor does not have this capability. The resulting speed-torque characteristics are shown in Figure 1.38. Due to the constant induction in the rotor, the AC asynchronous motor achieves relatively low efficiencies compared to the synchronous motor and therefore must at least be cooled passively.

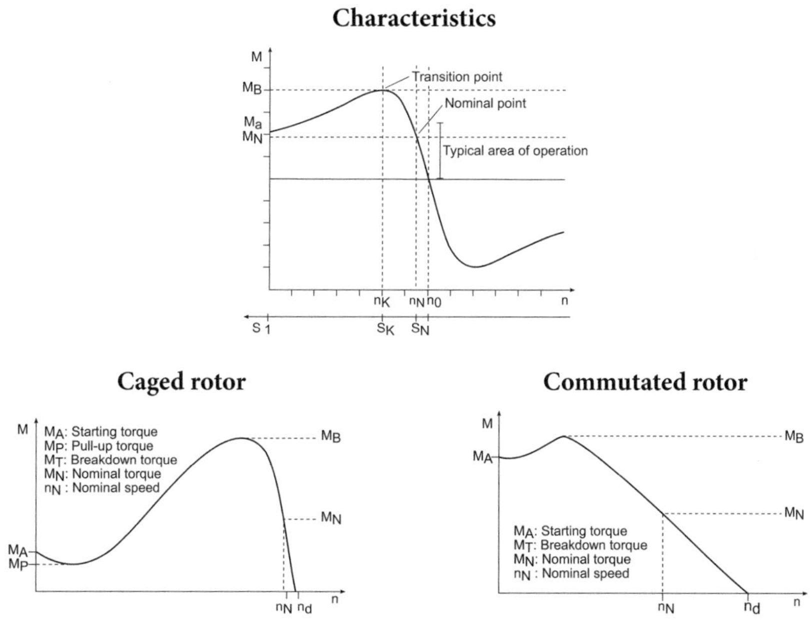

Figure 1.38 Characteristics of an asynchronous machine

Due to its simple structure and hence low costs it is still suitable for a number of applications. They range from industrial drives to traction drives, such as in trams or in elevators, to miniature motors in medical technology with speeds of 300,000 rpm and above.

Introduction to servo motor technology

In motor applications the servo motor type plays an important role, because it does not differentiate in terms of its electromagnetic structure, but in its mode of operation. The rotor position, rotational speed and acceleration can be determined and can therefore be controlled as well. Resolvers, incremental encoders or absolute encoders are used as sensors. A resolver consists of two coils in the stator arranged at a 90° angle, which are connected to the rotor either inductively or via the slip contacts. The resulting electrical signal from the phase-shift always provides the absolute value of the rotor position. Incremental encoders are usually sensors mounted directly on the shaft and determine the relative change in position using different procedures. The original position is not known without referencing or calibration. The absolute encoders also use different optical, magnetic or electrical procedures to determine the absolute position. These encoders are usually mounted directly on the shaft as well. Brushless DC motors are trending in servo motor technology with the use of resolvers, because they can meet the requirements of energy efficiency due to their simple structure.

Comparison of motor types

A comparison of motor properties can be made based on rotor properties due to similar stator design. The most common motor types, synchronous and asynchronous motors, can be found in Table 1.8 [Os-10].

Table 1.8 Comparison of different motor types based on their rotors according to [Os-10]

	ASM	PMSM (OM1)	PMSM (OM2)	PMSM (BM)	EESM
Costs	Low	Middle	High	High	Middle
Magnet	None	Yes	Yes	Yes	None
Additional Magnet fixture	None	None	Bandage/ gluing	None	None
Mass/Inertia	High	Low	Middle	High	High
Field weakening	Good	Poor	Poor	Middle	Good
Rotor losses	High	Very low	Low	Low	Middle

For the synchronous motor, different versions with surface mounted magnets excluding (OM1) and including overmolding (OM2) and buried magnets (BM) are compared. The first version can be designed either with or without bandage. The asynchronous motor is affordable due to its design without magnets, but possesses high masses with high losses due to its passive caged rotor. The resulting efficiency is worse compared to other types, but the field can be weakened for higher speeds, which is why the asynchronous motor is suitable for high speed applications. The design of synchronous motors with surface magnets and without bandage has the highest efficiency. The rotor mass is low as well. However, the rotors speeds are limited due to the lack of overmolding fixation. The transition from bandaged to buried magnets results in an efficiency reduction. But higher speeds are possible due to better field weakening properties.

Electromagnetic motor design is a well-researched field that provides extensive literature. Therefore, the focus of this section is on basic knowledge of the different motor technologies only to the extent that is relevant to winding technology. Typical applications for servo motors are in the area of automation, which includes winding technology. Asynchronous motors, on the other hand, are used for simple applications such as conveyor technology or fans due to their cost-efficient design. Brushless DC motors are increasingly used for the small power segments due to the wear friendly brushless design and the system efficiency, which is based on better control via the power electronics.

1.3.3 Applications of electric motors

Generally, according to [Gar-08] electric motor applications can be divided into four segments: household devices, automotive technology, consumer devices and industrial applications. Typical applications for different motor types are listed according to [Gar-08] and their respective segments. A variety of winding components used in cars are illustrated in Figure 1.39.

Figure 1.39 Variety of coils for automotive applications

Automotive technology: electrical traction drive, electrical windows, electrical central locking system, electrical sun roof, electrical antenna drive, electrical starter, electrical windshield wipers, electrical fuel pump, electrical oil pump, motor cooling fan, electronical stability control, anti-lock brakes, fresh air fan, throttle valve adjustment, electric seats, electric mirrors, compressor for air conditioning, etc.

Household devices: refrigerators, freezers, dishwashers, garbage chutes, garbage compactors, turntables in microwaves, washing machines, dryers, ventilators, compressors for air conditioning, evaporation coolers, chimney ventilators, air dryers, air humidifiers, exhaust fans, hairdryers, blenders, shredders, coffee machines, electric mixers, juicers, can openers, steam cleaners, vacuum cleaners, polishers, sewing machines, etc.

Consumer devices: computers, home trainers, document shredders, garage door openers, circulating pumps for swimming pools and ponds, sauna pumps, toys, radio-controlled cars and robots, fans for cooling electric devices, drives for hard disc drives, drives for cameras, drives for CD/DVD-drives, feed drives in fax machines and printers, lawn mowers, lawn trimmers, chain saws, power tools, etc.

Industrial applications: machines for food manufacturing, pumps, conveyor belts, traction drives, precision drives for robots, printing machines and tool machines, drives in medical technology, servo drives, drives for all areas of industrial manufacturing, such as textile machines, etc.

Based on the different applications, examples for some of the motor types are listed in Table 1.9.

Table 1.9 Selection of different drive applications depending in motor type according to [Gar-08]

Motor type \ Application	Heating & air conditioning	Automotive auxiliary drive	Lifting and material handling	White goods	Craftsman appliances	Office and computer technology	Servo drives for machine tools	Servo drives for handling tasks	Servo drives for packing technology Servo drives for handling tasks	Spindle drives for machining	Pumps and ventilators	Centrifuges and mixers	Clock drives	Testing technology & positioning
Step motor						x			x					x
Switched reluctance motor						x			x					x
Three-phase Synchronous machine			x				x	x	x	x	x	x		
Three-phase asynchronous machine			x				x	x	x	x	x	x		
One phase motor with caged rotor and capacitor aid	x			x							x	x		
One phase motor with magnet rotor and capacitor aid	x			x							x	x	x	
Universal motor				x	x									
Electronic (EC-) motor	x	x			x	x			x	x	x	x		
DC-motor series wound		x		x	x									
DC-motor shunt wound		x		x	x									
DC-motor permanently excited		x		x	x	x								x

The design of electric motors and the customisation of the desired characteristics, for instance power, speed or torque, as well as specifics such as cogging torque or torque ripples are provided by the product and taken into account by the product designer. Most relevant for winding technology is the geometry of rotor and stator, the groove geometries and the manufacturing design as full lamination stack or segmented single-teeth. The winding is the centerpiece of the motor. It is characterised by the winding scheme or coil type, as well as the number of turns and the wire gauge. Following overviews were created to provide a better comprehension of typical parameters with regard to the different motor applications; for the number of turns in Figure 1.40, and for the wire gauges in Figure 1.41.

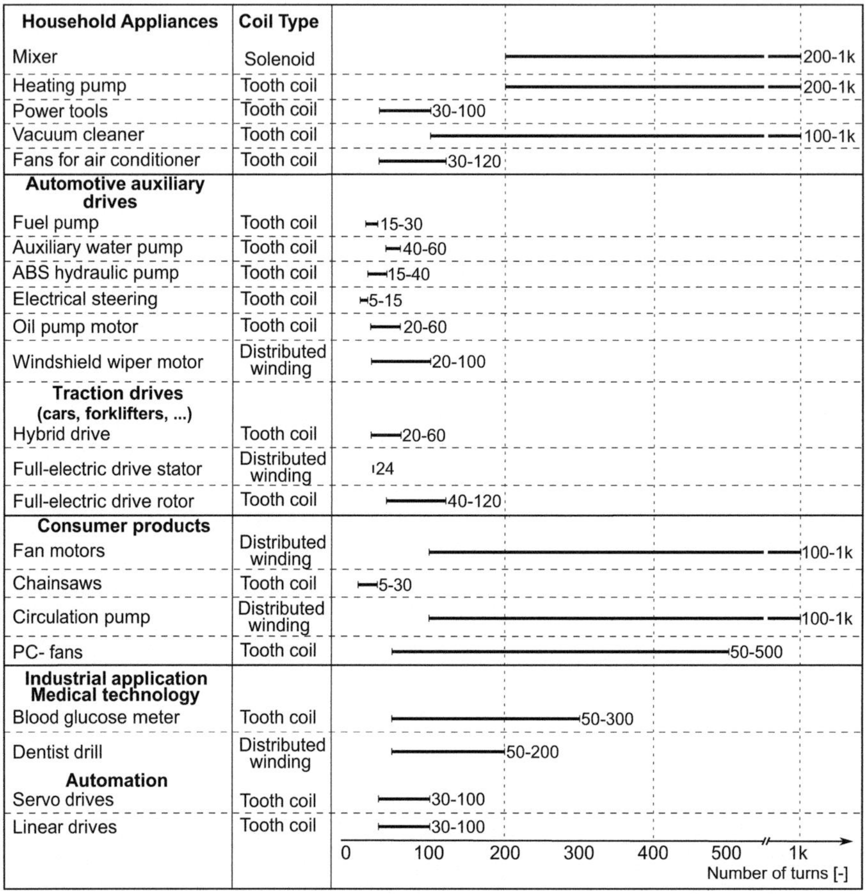

Figure 1.40 Number of turns for typical motor applications

The energy efficiency requirements for electric drives have lead to new challenges for the electromagnetic design and the manufacturing technologies. The main reasons for an increase in demand for efficient electric motors are stricter regulations by the legislation and the high distribution of electric motors. The high distribution results in large potentials for CO2 reduction or reduced energy consumption respectively. According to [Wai-11] 40 % of the world's electrical energy demand was used for electric motors in 2011, of which

9 % were accounted for in the small power segment (up to 0.75 kW), 68 % in the medium power segment (0.75 to 375 kW) and 23 % in high power segment with more than 375 kW. Based on these facts, the legal requirements for energy efficiency became stricter in 2009 and a new efficiency class IE-4 was introduced (Figure 1.32). Since 2015, all motors belonging to the medium power segment in the EU have to at least comply with the efficiency class IE-3. From 2017 onwards, this requirement is also valid for the high power segment.

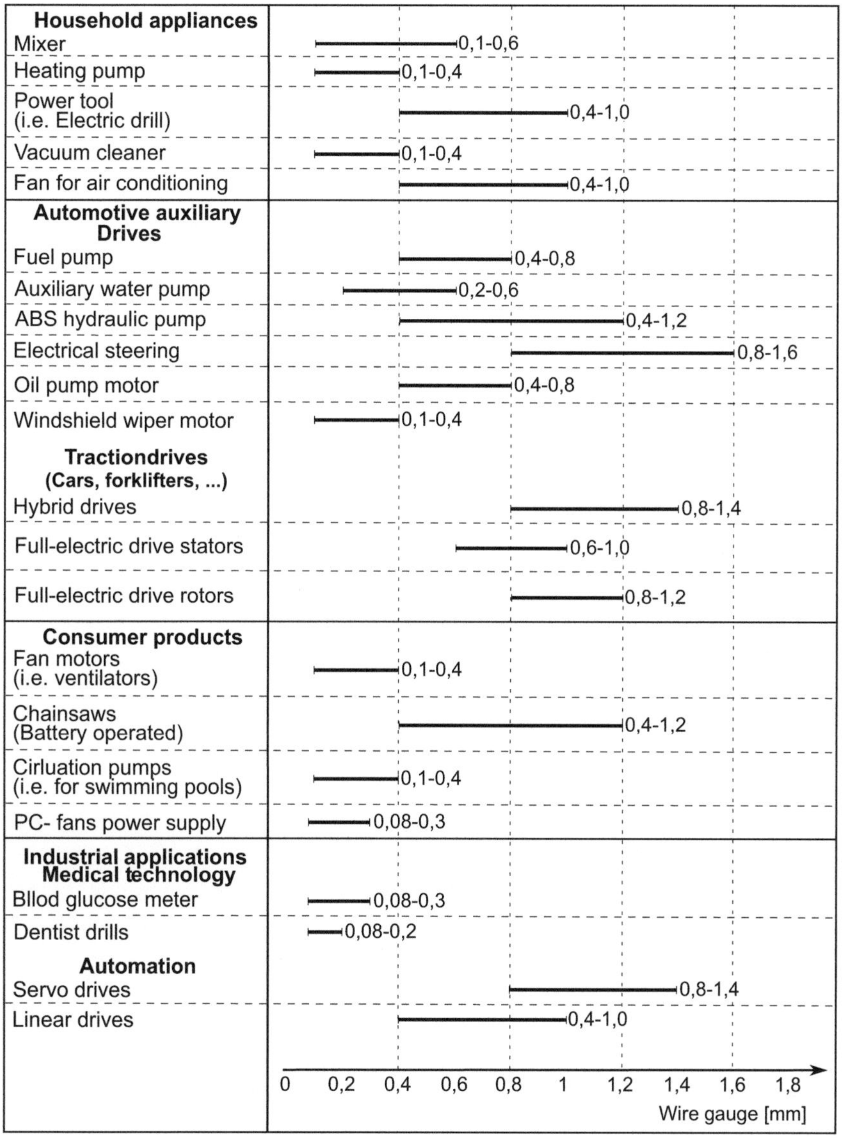

Figure 1.41 Wire gauges for typical motor applications

Following this trend the market share of efficient motors has been rising steadily since 1990 (Figure 1.42). However, to achieve the necessary improvements in electric motor technology, changes in design and control are needed, as well as more efficient manufacturing processes to provide optimised parts for electric motors. All motor types require an excitation through a winding in stator or rotor, which is why coil winding technology has assumed a key role in this context.

The increasing electrification in all areas of daily life will further increase the demand for electric motors. Especially the demand for high efficiencies motors, such as in the automotive drivetrain, is growing significantly. Consequently, more efficient windings are needed for all kinds of applications of electric motors.

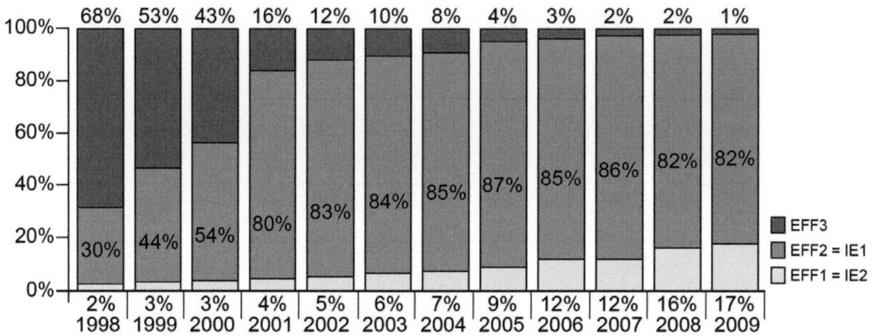

Figure 1.42 Distribution of efficiency classes

To meet these requirements, a design suitable for manufacturing is necessary and a continuous improvement of manufacturing processes is the objective of the manufacturing industry. The general requirements, which result from specific applications of electric motors and which are determined by the product designer when choosing a motor type, are numerous and must be considered during design. They are directly reflected in the requirements for production processes. This aspect is elaborated in the following section. The application areas of electric motors are numerous and, depending on their uses, comprise different challenges for winding technology. In terms of energy efficiency, the construction of the winding must be compact and feature the lowest possible electrical resistance. Due to the direct dependency of the conductor cross sectional area and the electrical resistance of a coil, there is a trend towards larger wire gauges and lower numbers of turns, or parallel wire strands respectively. For mobile applications in particular, there is a trend towards designs with higher exciter voltages and therefore larger wire cross sections and higher number of turns. Processing individual thicker wires represents a major challenge, as the positioning accuracy of individual wires during assembly has a higher influence on the winding properties.

1.3.4 General requirements for windings

The electric motor's requirements for its winding go beyond simply electromagnetic properties and, as previously mentioned, depend heavily on the application. A comparison of different requirements for electric motor windings is summarised in Table 1.10.

Table 1.10 Requirements for winding resulting from motor requirements

Requirement for electric motor	Resulting requirements for winding
- Long lifespan	- Faultless winding insulation
- Speed-torque-characteristic	- Fitting inductance and resistance
- Spatial limitations	- Compact winding structure
- Behaviour in error scenarios	- High voltage strength
- Temperature resistance	- Thermal durability of insulation
- Resistance against media (oil, water)	- Chemical durability of insulation
- Electromagnetic compatibility	- Reliable insulation
- Noise emission and engine smoothness	- Vibration resistance of winding
- Power consumption and power losses	- Fitting electric resistance
- Power density	- Small weight, compact design
- Small production costs	- Good processing capabilities

For traction motors used in hybrid motors or in fully electric drivetrains the power density and efficiency of the winding is crucial. Due to the limited automotive installation space, the motor must be as compact and light as possible. The currently immature battery technology in terms of electromobility can only ensure short distances in comparison to combustion engine drivetrains. This can partially be compensated by more efficient electric motors; however, these still need to meet the required driving dynamics. Typical requirements of most other auxiliary drives in cars apply for this automotive application, too. Vibrations from the suspension are transferred to the drives, and depending on the location of the drives, temperatures may range from -40 to 180 °C. Also, the drives are exposed to the road's media, such as water and oil. However, if the motor is supposed to be used in the oil circulation system, it must comply with certain cleanliness levels to avoid contaminating the circulation system. Chemical resistance of the motor's components is a key requirement for this.

In automation technology the boundary conditions are even more diverse. A pump for coolants or lubricants in a tooling machine must be able to operate with many foreign particles in the medium. These might not be in direct contact with motor components, but they may settle on the periphery. In contrast, a disc grinder must keep out foreign particles in the air by insulating the winding using casting or impregnation. In steel mills, motors in the conveyors are heated up to 300 °C by the heat radiated from the raw iron. In flour grounding mills, fine dust can develop, which leads to the highest requirements for the motor's tightness and particle resistance. By contrast, special requirements apply for paint spraying operations for reasons of explosion prevention.

In medical technology, the trend is going towards smaller actuators, which can be positioned closer to their place of operation thanks to miniaturisation. For these small drives the required mechanical power can only be achieved by extremely high speeds with several 100,000 rpm. Absolute cleanliness in the production of the motor and the biocompatibility of the materials used must also be considered. Higher voltages are often applied to increase the power density of high efficiency motors thus allowing the construction of a more compact design. This leads to an increase of the intermediate circuit voltage in the inverters and therefore greater requirements for the dielectric strength and partial discharge strength of the winding. The particular requirements for the wire insulation are discussed in Sections 2.3.1 and 2.3.2. Insulation procedures for additional insulation measures and materials, also referred to as secondary insulation, are introduced in Section 3.4.6.

1.3.5 Concentrated windings for electric motors

There are various ways to build a stator winding as illustrated in Figure 1.43. It can be built from single-poles or single-tooth windings mounted onto inner or outer grooves of the stator, or from a pole chain with wound and connected segments. In pole chains, either partial pole chains can be used for each phase, or a full pole chain including all of the motor phases. Each pole chain can be manufactured from separate segments, which then support each other. Finally, winding full stator lamination stacks is an option, however the limited accessibility of the grooves due to the small groove slots impedes the winding process.

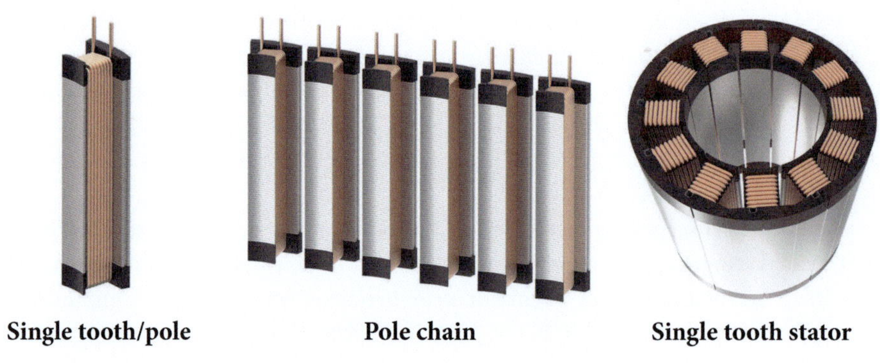

Figure 1.43 Stator winding and its elements

There are multiple ways of assembling the aforementioned individual segments to a stator. In the case of rotor windings for externally excited synchronous motors, a single-tooth design can be used. Alternatively, a distributed winding can also be wound onto a full lamination stack. Different variants are shown in Figure 1.44.

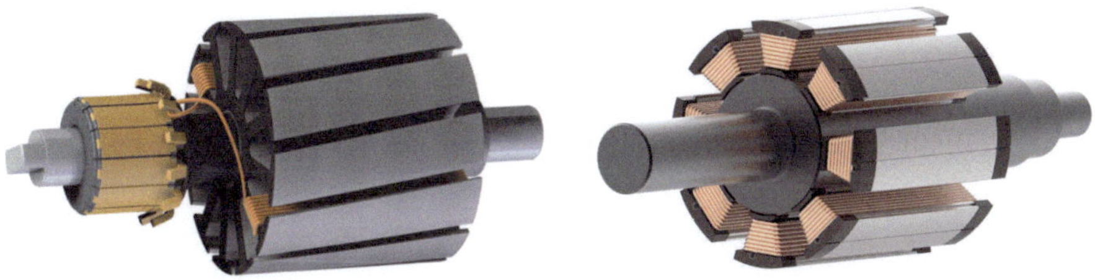

Commutated rotor with concentrated winding **Externally excited rotor with concentrated winding**

Figure 1.44 Examples for armature/rotor windings

The application requirements determine the motor design and type. Subsequently, the design of the coils inside the motor is a consequence thereof.

1.3.6 Distributed windings for electric motors

Generally, a distinction is made between distributed and concentrated windings. If only a partial single coil is wound around a bobbin or a tooth, it is referred to as concentrated winding. If the coil spreads over multiple teeth, it is referred to as distributed winding. The main parameters of distributed windings are:

- number of grooves N: resulting from the stator structure
- number of winding strands m: resulting from the number of phases
- step shortening v: resulting from a change in step width of a phase
- coil width w: resulting from the width of the coil between grooves
- pole count p: resulting from the number of magnetic poles

The number of holes q is an auxiliary unit for the design of electric motors and serves for the calculation of the magnetic field or the resulting motor torques. It is defined as:

$$q = \frac{N}{2 \cdot p \cdot m}. \tag{1.41}$$

If the number of holes is an integer, this is referred to as an integral slot winding, as the ratio of number of grooves for each pole and strand is integral. In the opposite case, this is referred to as fractional slot winding. Other characteristic quantities are the pole pitch τ_p, the groove pitch τ_n and the winding step y_N. These are defined as:

$$\tau_p = \frac{N}{2 \cdot p}, \tau_n = \frac{\pi \cdot d}{N}, \gamma_n = \frac{N}{2 \cdot p}, \text{for stranded windings } \gamma_n = \tau_p - \nu \quad (1.42)$$

A winding can be divided into multiple layers per groove. Typically, there are single-layer or double-layer windings. For the single-layer winding, each groove only contains one coil side. In addition, all coil diameters must equal the average pole pitch, as otherwise there will be an overlap with other coil strands [Fi-13]. The single-layer winding can be carried out in two variations:

- coils of the same width: the diameter of each coil equals the pole pitch ($w = \tau_p$)
- concentric coils: the coil diameter only equals the pole pitch on average (different diameters of individual coils).

Winding with concentric coils is popular because it is easier to manufacture in series production [Fi-13]. In double-layer windings each groove contains two coil sides of different coils. This enables so-called "chorded windings". Here, the coil width does not equal the pole pitch, but is lengthened or shortened. Due to the topology of the double-layer winding the other strands in the grooves are not affected. In comparison to single-layer windings, twice as many coil groups are created, as a coil pole is always made up of two coil groups with opposing winding directions. Due to this chording, the field characteristics are closer to a sinusoidal shape [Fi-13]. A structural comparison of single-layer and double-layer windings is shown in Figure 1.45.

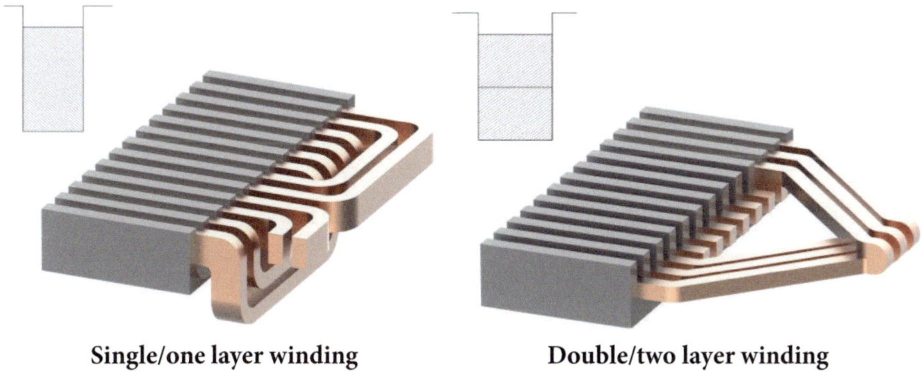

Figure 1.45 One and two layer windings according to [Mü-08]

For fractional groove windings the number of grooves for each pole and phase is a fractional number. This has positive consequences, because negatively affecting parts of the voltage are further suppressed due to the fractional groove winding as compared to an integral groove winding [Fi-13]. Therefore, it is used to suppress high harmonics, for example in large AC generators [Gie-03]. The design of electrical windings can be optimised using different criteria, such as the start-up behaviour, torque ripples or the maximum torque.

Consequently, numerous winding schemes exist for various applications. An extract is given in Figure 1.47.

For a better comprehension the lap and wave winding are explained in more detail. A comparing illustration is shown in Figure 1.46.

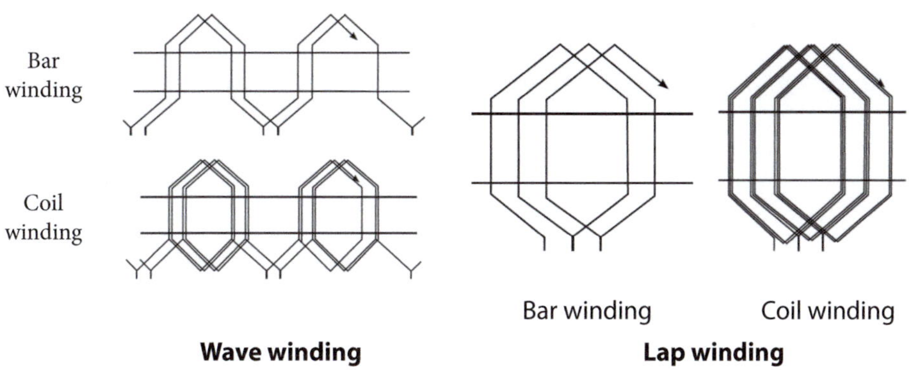

Figure 1.46 Comparison of wave and lap winding according to [Mü-08]

Thereby, it can be differentiated between structures made of conductor bars or diamond coils (Section 1.2.1), also referred to as hairpins, or a structure of wound coils. The number of lines in the scheme displays the number of conductors.

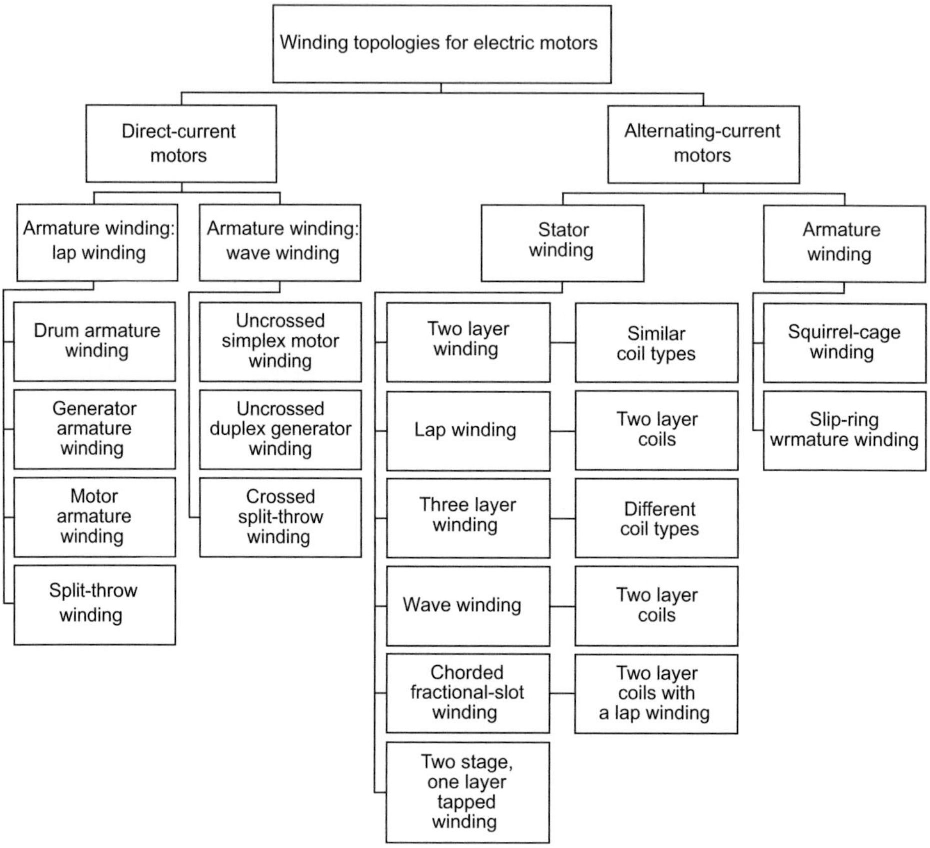

Figure 1.47 Overview of different winding topologies

For the lap winding, the end of the previous coil is connected to the beginning of the closest neighbouring coil. For the wave winding, the end of the previous coil is connected to the beginning of the next pole pair at the same point. By doing this, the number of the coils p is reached after one revolution. Each coil has a defined number of turns for each winding type. The resulting scheme for a distributed winding of a DC motor with four poles and 16 fins or bars is shown in Figure 1.48.

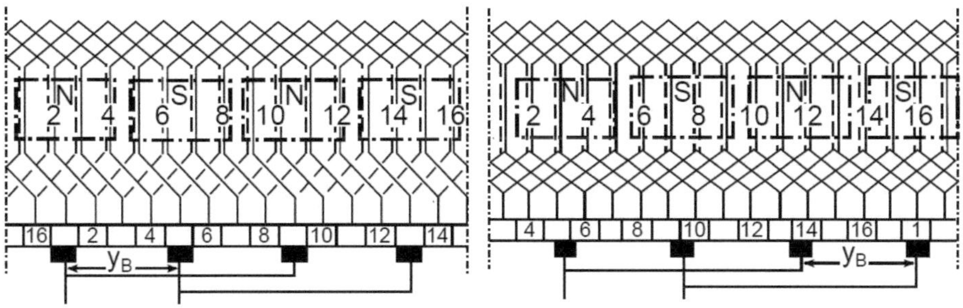

Lap winding **Wave winding**

Figure 1.48 Complete winding scheme of a wave winding and lap winding for DC-machine according to [Fi-06]

Extensive literature is available for electromagnetic winding design, for instance [Bal-69] and other publications. The winding manufacturing was first mentioned in a book by Tzscheutschler [Tzs-90]. Usually, only multi-stage production procedures can be used for such complex structures. Typically, the winding is aligned on a tool and then assembled separately outside the stator. However, there is a trend towards direct winding of the stator. The major challenge of this technological approach is the accurate placement of the connecting wires. End disk or end cap design becomes even more complex due to the spatial requirements for the motor. The minimal bending radii of the wire, in particular, must be taken into account during placing at the end disks or end caps. For thicker wire gauges the mechanical stability of the cap or bars, has to be ensured for the placement, using defined minimal wall stiffness and thicknesses respectively.

Typical stator windings are distributed windings, which are manufactured using the insert technology (Section 3.3.8). In contrast, due to possible economies of scales, in some product segments there is a trend towards manufacturing single-teeth. However, these are limited to their application areas. For starter generators there is a trend towards diamond coils or towards wave winding.

References

[Alt-08] Alt H: Dualität: Mechanik – Elektrik. Lecture notes, FH Aachen

[Bal-69] Bala C, Fetita A, Lefter V: Handbuch der Wickeltechnik elektrischer Maschinen. Verlag Technik 1969

[BASF-14] BASF: Reibung und Verschleiß bei Polymerwerkstoffen. Technical brochure

[Br-12] Braeuer P et al.: 3D Screen Printing technology – Opportunities to use revolutionary materials and machine designs. 2nd International Electric Drives Production Conference (EDPC) 2012. ISBN 978-1-4673-3007-7

[Fi-13] Fischer R: Elektrische Maschinen. Carl Hanser Verlag, München 2013. ISBN 978-3446438132

[Gar-08] Garbrecht FW: Auswahl von Elektromotoren leicht gemacht: Der Weg von der Anwendungsanalyse zum richtig dimensionierten Elektromotor. 1st ed., VDE Verlag 2008. ISBN 978-3800729111

[Gi-03] Giersch H-U, Harthus H, Vogelsang N: Elektrische Maschinen. Springer Verlag 2003. ISBN 3-519-46821-2

[Gro-14] Groninger M et al.: Cast coils for electrical machines and their application in automotive and industrial drive systems. 4th International Electric Drives Production Conference (EDPC) 2014. ISBN 978-1-4799-5008-9

[Hi-02] Hilleringmann U: Silizium-Halbleitertechnologie. eBook, Vieweg + Teubner 2002. ISBN 978-3-322-94119-0

[HUG 15] INDUSTRIETECHNIKKATALOG 2015 der Firma HUG Industrietechnik. URL: http://www.hug-technik.com/inhalt/ta/werkstoff_kunststoffe.htm

[Mü-08] Müller, Vogt, Ponick: Berechnung elektrischer Maschinen. Wiley-VCH 2008

[Os-10] Oswald J: Vor- und Nachteile verschiedener Motorkonzepte für Fahrantriebe. DRIVE-E Akademie 2010

[Reh-03] Rehrman J: Das InterNetzteil- und Konverter-Handbuch. URL: http://www.joretronik.de/Web_NT_Buch/Vorwort/Vorwort.html (7.8.15)

[Ru-08] Rudolph M, Wagner U: Energieanwendungstechnik. VDI/Springer 2008. ISBN 978-3-540-79021-1

[Tzs-90] Tzscheutschler R, Olbrisch H, Jordan W: Technologie des Elektromaschinenbaus. Verlag Technik 1990. ISBN-10: 3341008519

VDI-16 VDI-Richtlinie: VDI/VDE 2616 Blatt 1 & 2: Härteprüfung an metallischen Werkstoffen, 2014-07, ICS-Nummer: 83.060

[Wai-11] Waide P, Brunner C: Energy-Efficiency Policy Opportunities for Electric Motor-Driven Systems (Working Paper). International Energy Agency 2011

[Wic-00] Wicki S: Berechnung der Spulen. Project Report for Alpha Step, FH Aargau 10.07.2000

Enamelled copper wire

2

The purpose of this chapter is to provide the reader with knowledge about the properties and the manufacturing process of enamelled copper wire. Typical materials and geometries of the wire cross section (wire profile) as well as alternative variants such as stranded wire will be introduced. This is followed by a detailed presentation of the process chain for manufacturing enamelled wire.

2.1 Conductor properties

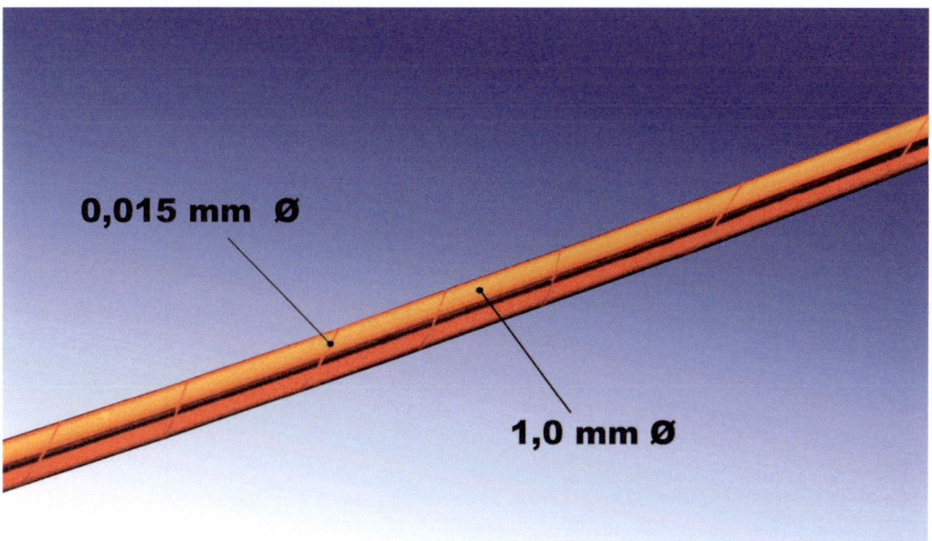

Figure 2.1 Dimensions of magnet wire

This section deals with copper as a conductor material and introduces other alternative materials. Besides the electrical properties, possible influences on the mechanical and thermal properties of copper are discussed. For the use of enamelled copper wire the standardised processing properties according to DIN 60317 are introduced. Besides the material, different geometries are discussed, with aspects ranging from design support of stranded wire to the use of profile wire.

Consequently, this section is directed at readers, who are interested in the properties of magnet wires, whether from a manufacturing standpoint or the development of winding goods.

2.1.1 Copper as an electrical conductor

Due to its outstanding electrical conducting capabilities of $\kappa_{Cu} = 60\ \mathrm{m/(\Omega \cdot mm^2)}$, copper is widely used in electrical engineering. Only silver has a better conductivity of around 6 %, although it has significantly higher raw material costs. To achieve such high conductivity values, copper has to have the lowest contamination possible. To ensure the minimum conductivity of $\kappa_{Cu,min} = 58\ \mathrm{m/(\Omega \cdot mm^2)}$ required according to the ISO standards for wire manufacturing, electrolytic copper with a purity of at least 99.9 % has to be used [DKI-00]. Several typical material parameters are summarised in Table 2.1. To simplify comparison, the values were normed to copper values in %.

Table 2.1 Characteristic metal values and prices (effective 06.2015)

Metal	Rel. electric conductivity	Rel. thermal conductivity	Rel. thermal elongation	Normalized density	Normalized costs
Silver	106	108	112	118	8,917
Copper	**100**	**100**	**100**	**100**	**100**
Gold	72	76	100	217	633,333
Aluminium	62	56	135	30	28
Magnesium	39	41	153	19	42
Zinc	29	29	176	80	36
Nickel	25	15	76	100	208
Cobalt	18	17	71	100	550
Steel	13...17	13...17	71	82	10
Platinum	16	18	53	242	595,000
Tin	15	17	124	82	258
Lead	8	9	165	127	32

For the processing of copper as magnet wire there are different international standards in terms of material properties. DIN EN 1977 describes *Copper and copper alloys – Copper drawing stock (wire rod)*, a semi-finished product for wire manufacturing. DIN EN 13601 describes *Copper and copper alloys – Copper rod, bar and wire for general electrical purposes.* DIN EN 13602 specifically describes the properties of *Copper and copper alloys – Drawn, round copper wire for the manufacture of electrical conductors*, which is complemented by DIN EN 13605 *Copper and copper alloys – Copper profiles and profiled wire for electrical purposes.*

Characteristics of standard winding-wires

The aim of this section is to introduce a number of key properties for standard magnet wires. Besides the electrical properties, other application-relevant properties of standard magnet wire, like geometry and minimum elongation at break, are defined in DIN EN 60317. The maximum allowed deviation of the wire's outer gauge, depending on the bare wire gauge and the insulation thickness from grade 1 to grade 3, is shown in Figure 2.2. It becomes clear that the different insulation grades have similar deviations in diameter. However, these are considerably greater than the deviations of bare wire for a range of up to 4 mm.

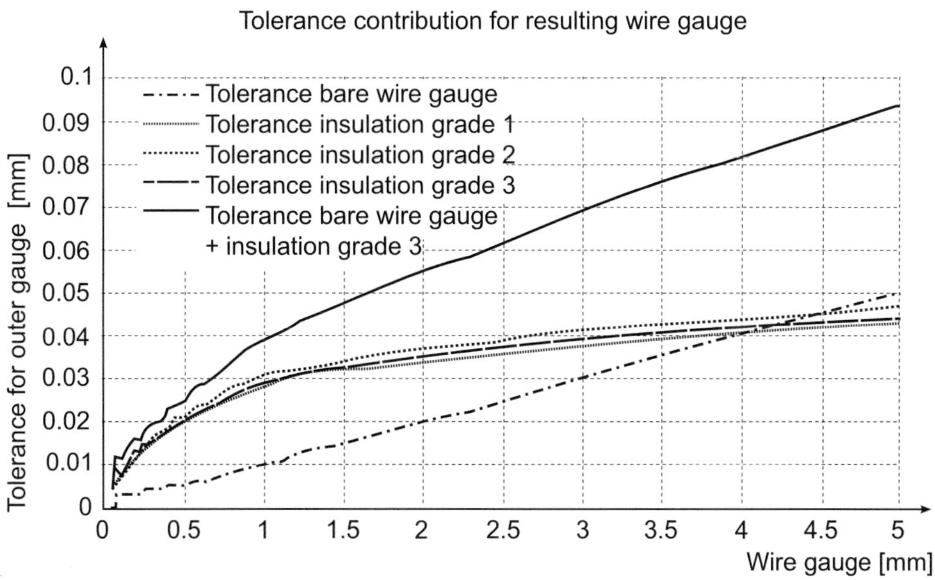

Figure 2.2 Influence of tolerances on wire gauge

The resulting mechanical fill factor, which is defined as the relation of wire cross section to the whole winding space, is illustrated in Figure 2.3 for different numbers of layers. The figure shows that small changes in the wire gauge have a big impact on the achievable fill factor.

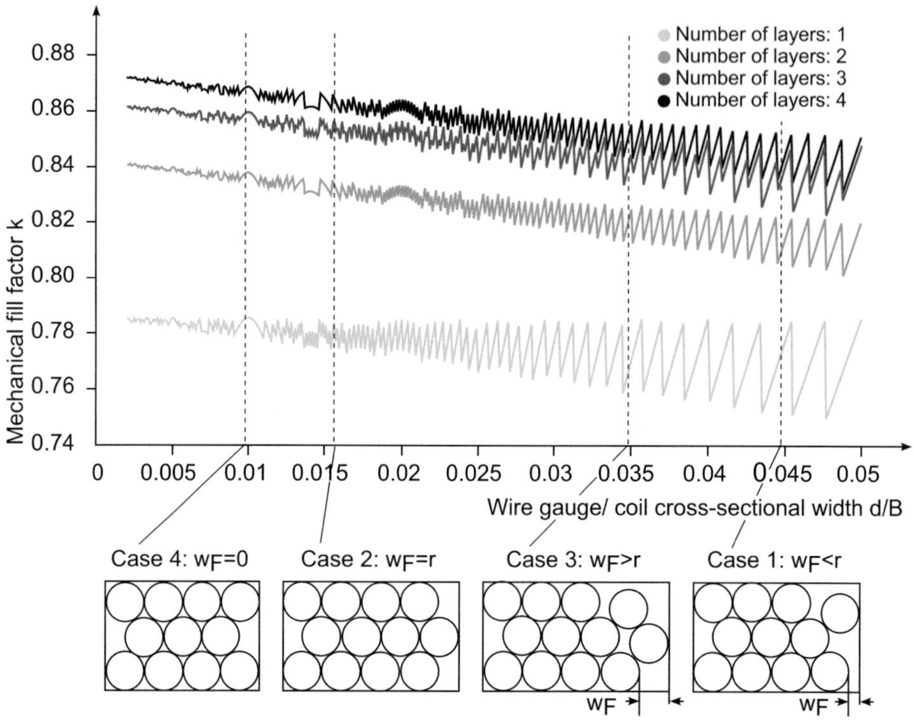

Figure 2.3 Influence of wire gauge and winding width on fill factor

Additionally, further mechanical quantities, which are relevant for the processing, like spring-back or elongation at break, are given in DIN EN 60317 (Figure 2.4). For larger wire gauges the elongation at break, a measure of maximum wire deformation, increases, while the allowed angle of spring-back decreases. Consequently, thicker wires can be strained more and feature lower spring-back angles. The required bending force, however, increases.

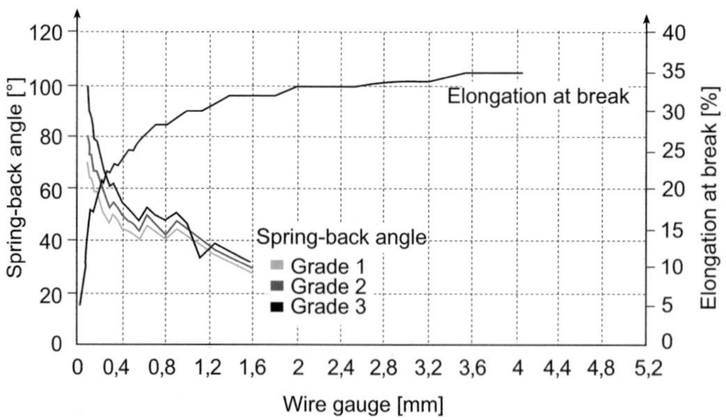

Figure 2.4 Spring-back behaviour and elongation at break for different wire gauges

When using standard magnet wire, is it important to know that the insulation is allowed to contain a standardised number of imperfections in terms of closeness. The number depends on the wire gauge and the chosen insulation thickness according to Table 2.2:

Table 2.2 Allowed number of faulty insulation spots on the wire surface according to EN 60317-0-1:2008 (p. 18)

Nominal gauge in mm		**Max. amount of local failures in 30 m**		
Above	Until	Grade 1/1B	Grade 2/2B	Grade 3
–	0.05	60	24	–
0.05	0.08	60	24	3
0.08	0.125	40	15	3
0.125	1.6	25	5	3

Electrical and thermal properties of copper

The specific electrical resistance of materials can generally be determined by the Matthiessen rule. This rule describes the dependency of the specific electrical resistance on the temperature and the material composition, for example, lattice defects in the macroscopic structure or foreign atoms through alloying of the material. The dependency of the specific electrical conductivity on the alloying with other materials or the temperature is shown in Figure 2.5.

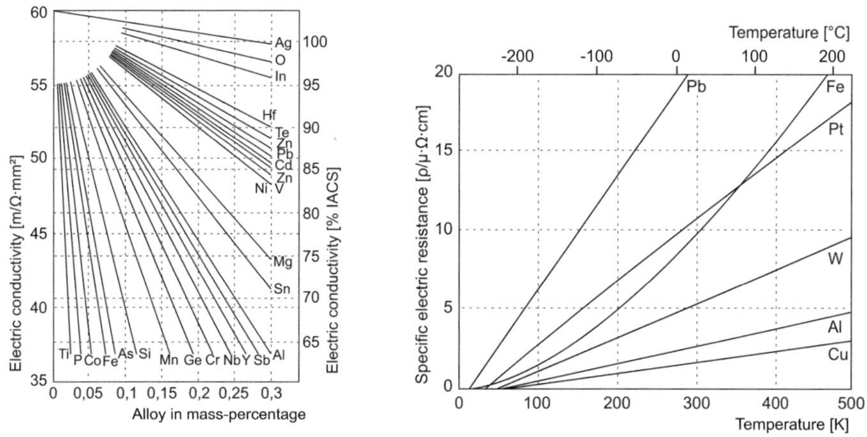

Figure 2.5 Influences on electric conductivity according to [DKI-00, p. 7], [Tif-07, p. 70]

Electrical conductivity is defined as the reciprocal of the electrical resistance. However, the electrical conductivity of copper wire is not only influenced by the alloying with other metals, but also by the processing. Processing with repeated bending deformation may cause strain hardening, which reduces the electrical conductivity. This issue is covered in more detail in Section 3.1.1. A similar law is valid for thermal conductivity properties ascribing very good conductivity properties to copper, which are only surpassed by silver. Thermal conductivity properties are especially relevant for the thermal design of sub-assemblies. Electrical losses of coils, with the exception of reactive power of inductive components, are entirely converted into heat. This is why coils heat up significantly during operation, in particular, with higher loads. As a result, a good thermal conductivity of the coil wire used is a prerequisite to prevent the component from overheating. Due to the increasing dependency of the electrical resistance on the temperature, the electrical losses increase with higher conductor temperatures. This is especially critical for application with a constant high current. Therefore, an effective cooling of the component or sub-assembly is necessary for high efficiency. The thermal longitudinal growth is small for copper compared to other metals, enabling dimensional accuracy of the sub-assembly even at high operating temperatures. This permits applications of electric motors with operating conditions in ambient temperatures from -80 °C to 300 °C [Dem-11].

Mechanical properties of copper

Besides the positive electrical properties, the wide distribution of copper in electrical engineering and other technical areas can be traced back to the good mechanical properties, especially in terms of processing. Copper is a very ductile material and can therefore be deformed easily even in a cold state. A measure for reshaping capabilities of materials is a high degree of deformation, also called forming capacity. It is very high for copper and can be displayed using an example from wire manufacturing. During production, the wire cross section can be reduced by more than 99 % by continuously drawing the wire, without causing damage to the material [DKI-00]. As opposed to other materials, copper does not have a fixed yield limit, which means that there is no fixed point for the transition from elastic to a plastic deformation. Typically, a value of 0.2 % of the creep limit is defined as the yield limit. For copper this ranges from 40 N/mm2 to 80 N/mm2. The cross-dependencies of hardness, yield limit and elongation on deformation or load are shown in Figure 2.6. A distinct feature of copper is that deformations of up to 10 % have a relatively low impact on the mechanical processing properties by forming.

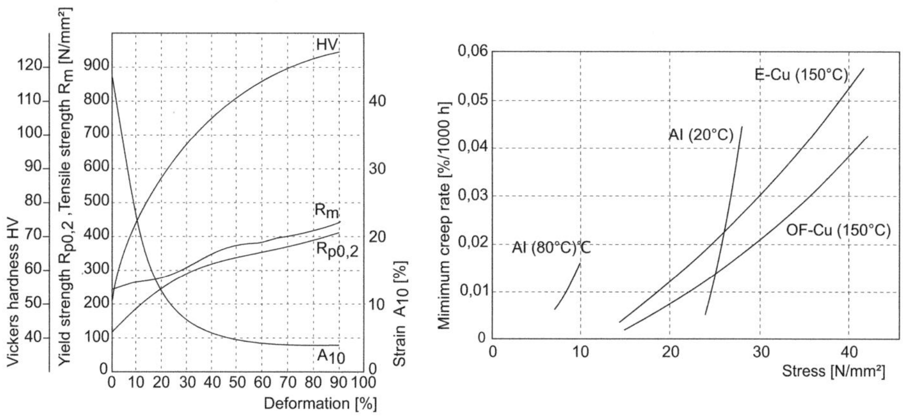

Figure 2.6 Influences on mechanical characteristics according to [DKI-00, p. 9]

Besides strength to single mechanical stresses, for instance during processing, one of the advantages of copper is its high creep strength and a creep velocity tending towards zero. This describes the material's elongation under constant stress. Creep strength is especially relevant for the manufacturing of electric motors, in which the winding as a sub-assembly takes on electrical and mechanical functionalities. Typical processing properties of different strength classes of a certain copper material (Cu-ETP) are summarised in Table 2.3.

Table 2.3 Processing characteristics of Cu-ETP [DKI-00]

Strength class	Electric conductivity in m/Ωmm²	Tensile strength R_m in N/mm²	0,2 % yield strength $R_{p0,2}$ in N/mm²	Brinell hardness HV
R200	58	200-250	max. 100	55
R220	58	220-260	max. 140	55
R240	57	240-300	max. 180	80
R290	57	290-360	max. 250	95
R360	56	min. 360	320	min 110

Generally, a higher mechanical strength of copper leads to slightly inferior electrical properties. Moreover, wires with higher strength can endure higher stresses in the process due to their robustness, but have smaller forming capabilities. Processing of wires with higher stresses, for example at higher process speeds, is usually associated with poorer product features of the winding good, like a higher electrical resistance or a poor final contour of the wire. A detailed explanation can be found in Section 3.1.1. The underlying physical effects are elaborated in [Kug-09].

Chemical properties of copper

One of the chemical properties of copper is the good corrosion resistance. This is due to the development of an oxide layer on the surface, which protects the underlying layers from reactive media. This effect can be seen in old copper roofs, where exposure to the weather leads to the development of a green oxide layer. However, this can be prevented by the influence of complex salts. Consequently, there is a risk of completely corroding the copper conductor. Additionally, copper has further properties, for instance an anti-bacterial effect, although these are not discussed further in this book. Further information can be acquired through the German Copper Institute or the Copper Alliance [DKI-00].

The use of copper in electrical engineering is based on the relation of costs to specific electric conductivity. The good forming properties of the material enable good processing characteristics. These are the reasons for the current dominance of copper as a conductor material for the different type of coils.

2.1.2 Other conductor materials

Depending on the application, besides other pure materials, different combinations of conductor materials may be used.

Table 2.4 Comparison: Aluminium, copper and copper-clad aluminium according to [Bec-13, p. 8]

	Copper	**Aluminium**	**CCA (10 %)**
Appearance on earth	0,01 %	7,75 %	-
Density	8,92 g/m³	2,70 g/m³	3,63 g/m³
Melting point	1083 °C	658 °C	658 °C
Thermal expansion	$17{,}0 \times 10^{-6}$ 1/K	$23{,}5 \times 10^{-6}$ 1/K	-
Thermal conductivity	385 W/mK	230 W/mK	240 W/mK
Spec. Thermal capacity	382 J/(kg K)	896 J/(kg K)	862 J/(kg K)
Young`s Modulus	$13{,}0 \times 10^4$ MPa	$7{,}1 \times 10^4$ MPa	-
Electric conductivity DC	> 58,58 m/(Ω mm²)	> 35,67 m/(Ω mm²)	> 37,35 m/(Ω mm²)*
El. conductivity AC (> 5 MHz)	> 58,58 m/(Ω mm²)	> 35,67 m/(Ω mm²)	> 58,58 m/(Ω mm²)
Costs (2014)	5,741 €/kg	1,416 €/kg	3,6 €/kg – 13,7 €/kg
Weldability	Good	Poor	Good
Oxidation rate	Small	High	Small
Formability	Good	– Bending strength – Long term stability	– Bending strength

Pure aluminium is the most common alternative to copper, resulting in a trend towards copper coated aluminium, so-called copper-clad aluminium. For higher frequencies the penetration depth of the current into the conductor decreases due to the skin effect (Section 1.2.2). Consequently, the conductivity in the outer conductor radius is important and can be enhanced using the copper coating. The resulting properties in comparison to pure aluminium and copper conductors are shown in Table 2.4.

Besides enhanced conductivity, the benefit of this coating lies in the lower material costs of aluminium as carrier material. However, as opposed to copper it cannot be processed as easily. Contacting, i.e. by welding, is difficult or impossible, and the allowed mechanical stresses are significantly lower compared to copper. The thermal properties are mixed. The conductivity is generally lower, although the heat capacity is higher. This means more energy is needed to heat up an aluminium conductor than a copper conductor. However, the warmed aluminium conductor conducts its heat slower to its surroundings. The choice of conductor material therefore always has an influence on the product and manufacturing properties. Although a product's economic efficiency can be increased by the use of lower material prices, it is possible that this implies changes to other product parameters due to changed material properties. As an example for design, different bobbin radii or their edge radii can be stated.

Table 2.5 Selected application scenarios for different wire materials

Wire material	Application	Technical Background
Aluminium	Speaker, microphones, ballast coils (neon-lights), chokes (transformers), clutch coils (air condition compressor)	- Weight reduction - Material costs
Brass	Antennas, hollow conductors, electronics for coils with thin wire diameters	- Good strength
Silver	Motors for medical applications, optical industries	- Good strength - Good spring characteristics - Electric conductivity
Gold	Antennas, Flip-Chip- and Chip-to-Chip-Bonding	- Suitable for good contacting quality
Nickel	Heating applications, electronics, chemical and electronic industry	- Good mechanical characteristics - Corrosion resistance - High temperature coefficient - High spec. electrical resistance
Special wire: CCA	High frequency coils	- High electric conductivity with low weight

Wire material	Application	Technical Background
Special wire: Constantan (55 % Ni 44 % Cr 1 % Mn)	Relays for motor protection switches	– Temperature independent resistances
Special wire: Nickel-Chrome	Heating resistance for the automobile industry	– Temperature resistant – Chemically durable
Special wire: Stainless steel	Welding wire	– Suitable for contacting

For the winding process, a higher material strength may lead to greater wire stress and thus to premature material failure. Terminating coated wire is difficult, as the coating's functionality must be ensured and it must not be damaged. Aside from welding, further contacting processes are available, which depend on the chosen material of the joining partners (Section 3.4.2). Generally, the subsequent manufacturing processes must be considered when choosing the material. To increase the strength of copper wire often silver is alloyed to it. Although silver has a better electrical conductivity than copper, the overall conductivity of the alloy decreases due to formation of a mixed crystal. Compared to other alloy materials, the use of silver displays the smallest reduction in electrical conductivity. In addition, the alloy improves the creep behaviour due to the higher strength. Creep behaviour describes the enduring of a material stress without breaking for a defined period of time. The use of further conductor materials usually originates from the application. Several examples are given in Table 2.5.

The properties relevant for processing several materials are summarised in Table 2.6. These can vary significantly depending on their manufacturing characteristics. Increasing the wire hardness leads to a lower yield limit. The elongation at break, or fracture strain, for a material defines the deformation capabilities of materials, and, for copper, it is associated directly with the level of re-crystallisation. It is determined as part of the tensile test and gives the relation of the change in length to the original length of the wire at the point of malfunction through breakage. It is therefore a key unit for winding technology and describes the maximum deformation. Typical values are between 25 % for 0.2 mm thick wires and 45 % for 4.1 mm thick wires. With each forming process the elongation at break is reduced due to the structural changes. For heavily deformed wires it can take values lower than 1 %. [Syn-15]

Table 2.6 Processing characteristics for different materials [Ele-15]

Material	Young's Modulus in kN/mm^2	Tensile strength R_m in N/mm^2	Yield strength $R_{p0,2}$ in N/mm^2	Maximum elongation at break
Copper	120	220...270	120...180	42 %
Aluminium	67	120...140	20	35 %
Brass	80...120	340...390	120...420	30 %
Silver	80	170...220	170	30 %
Gold	80	120...220	205	50 %
Platinum	170	150...380	185	39 %
Nickel	205	450...550	230...460	45 %
Stainless steel	200	540...850	230	45 %

It should be noted that for the winding process, materials with a high elastic modulus and at the same time high tensile or breaking strength in the elastic region, can be processed reliably. For applications in winding technology the yield limit is relevant as a process limit. It describes the necessary tensile force per wire cross sectional area, from which the wire is no longer elastically, but plastically deformed. Accordingly, this tensile force should not be exceeded since, aside from a permanent reduction of the wire gauge, strain hardening sets in. Both have negative influences on the electrical resistance. The hardening also leads to a reduction of the vibration resistance. Consequently, different minimum bending radii result, which must be considered in wire guidance or bobbin design. Plastic deformation, however, still takes place. The tensile strength describes the force per wire cross sectional area, which is required for a wire tear. This parameter is relevant for the defined separation, like tool-less tearing of the wire on the bobbin, or the separation of start and end wire.

The requirements arising from the use of the product define the choice of conductor material. After copper, aluminium is used often due to its low raw material costs with relatively good conductor properties. Nickel-chrome wire is used for the manufacturing of resistive components.

2.1.3 Geometries of the conductor cross section and conductor variants

Structure of a conductor

Generally, the structure of electrical conductors for winding technology is similar and, besides the material, only differs by its geometries. In the core of the conductor is the base material, which, as discussed in the previous sections, may consist of different materials.

A primer is applied to the base material to ensure the flawless application of the insulation layer. The overcoat is applied after the primer. Subsequently, the insulation is complemented by functional layers, such as self-bond enamel or a lubricant layer. The resulting wire structure is shown in Figure 2.7.

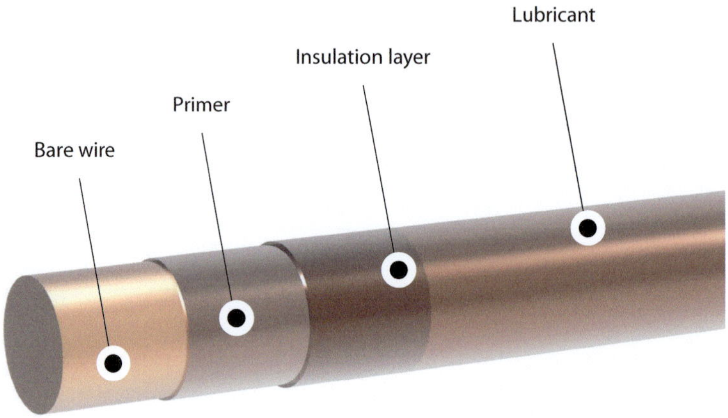

Figure 2.7 Structure of a conductor

The conductors can have different cross sectional shapes and different gauges. The most commonly used, besides circular cross sections, are conductors with rectangular cross sections. To fill a given winding space, different arrangements for both circular and rectangular cross sections are possible. Figure 2.8 illustrates different geometries and arrangements for winding stator slots.

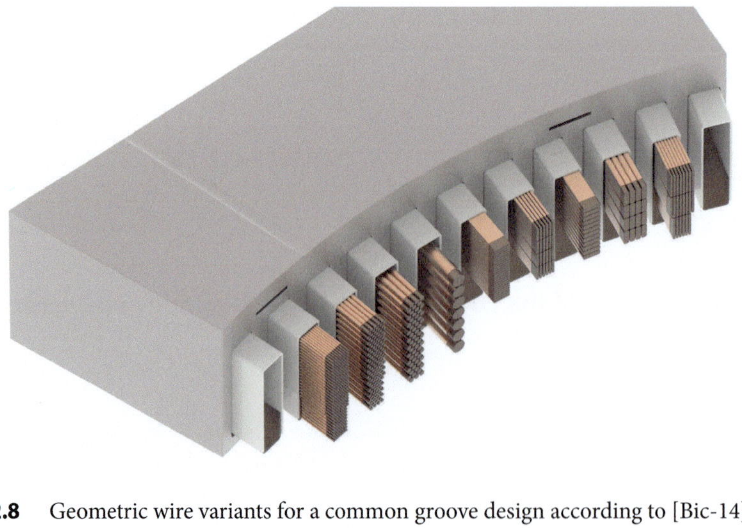

Figure 2.8 Geometric wire variants for a common groove design according to [Bic-14]

Stranded conductors

Besides the varying conductor geometry, small strands can be merged together into different bundles of conductors. For a circular conductor, this results in a so-called stranded conductor, for which different wires are twisted against each other.

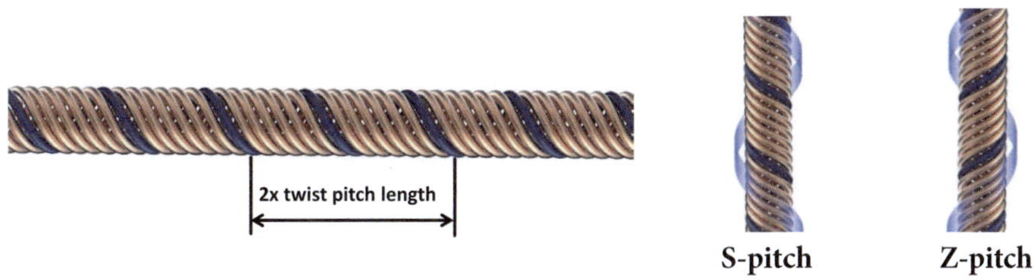

Figure 2.9 Characteristic parameters for stranded wires

This results in the so-called twist pitch. The twist pitch can be varied in pitch length and pitch direction; these are compared in Figure 2.9. To ensure dimensional accuracy, the individual enamelled copper wires (a) (Section 2.10) can be combined to strands without (b) and with overcoat (c). By using different strands, this can result in circular conductor profiles as well as quadratic (d) or rectangular cross sections (e) to increase the packing density in the winding space. Twisting the conductors offers electrical advantages. First of all, the influence of the skin effect is reduced due to the smaller conductor cross sections (Section 1.2.2). The same applies to the proximity effect, because due to the changing position of wires to each other, a uniform suppression of current between the insulated conductors cannot be established. The disadvantage of stranded conductors lies in the high connecting efforts of the parallel strands and the reduction of allowed conductor stresses in the winding process due to the reduced cross sections of the strands.

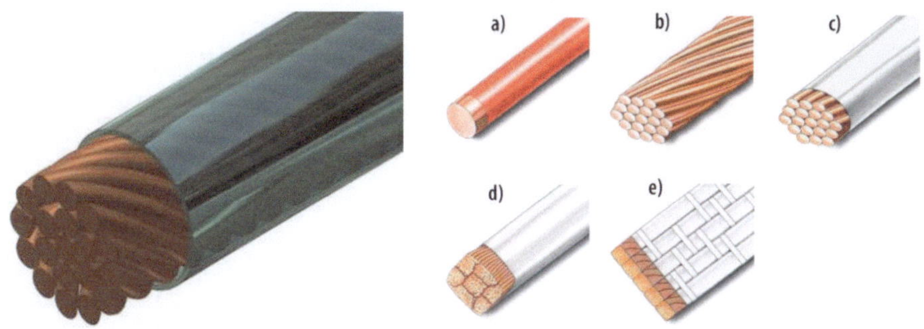

Figure 2.10 Stranded wires: Inner structure and outer contour
(Picture: Elektrisola GmbH & Co. KG, Rudolf Pack GmbH & Co. KG)

Depending on the number of stranding steps in manufacturing, one step can result in one bundle (Figure 2.11), or multiple stranding steps in multiple bundles. In terms of structure, a distinction is made between bunched stands and concentrically bunched strands. In bunched strands the individual conductors have a random position within the cross section. In contrast, the concentric structure leads to a ring-shaped and homogenous structure. Random arrangements can be chosen for a multi-bundle structure; these result from the requirements of the end product or the winding. According to publications by [Pac-10] as part of a research project, the systematic definition of the strand structure can reduce winding losses, for example in power electronics applications.

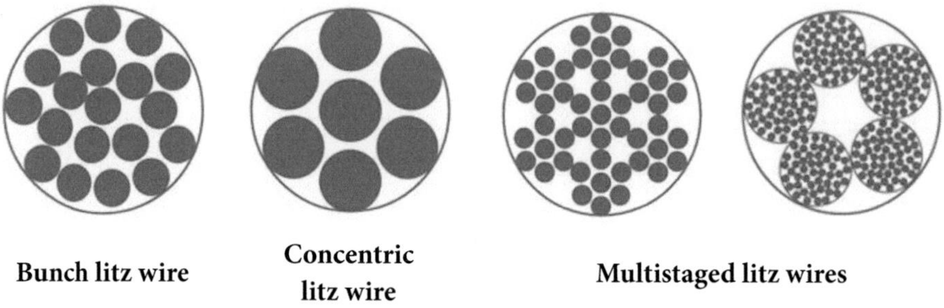

Figure 2.11 Structure variants of litz wires (Picture: Elektrisola GmbH & Co. KG)

For the design of stranded wire multiple aspects have to be considered. An overview of the relations between construction parameters of strands and the resulting strand properties is given in Table 2.7. The most important strand properties are the DC resistance R_{DC} and AC resistance R_{AC}, the current density J, the fill ratio of strand material to the available room $A_{Conductor}/A_{Strand}$, the breakdown voltage U_{BV}, as well as several chosen mechanical parameters.

Table 2.7 Design matrix for litz wire development according to [Ele-15a]

	Design parameters						
Litz wire characteristics	Conductor material	Conductor material	Number of wires	Number of strands	Twist pitch length	Twist pitch direction	Increase of varnish
R_{DC}	x	x	x				
R_{AC}	x	x		x	x		
J		x	x				
$A_{Conductor}/A_{Strand}$		x		x	x	x	x
U_{BV}							x
Outer diameter	x	x	x	x	x	x	x
Form stability		x	x	x	x	x	
Mech. flexibility	x	x	x	x	x		
Flexural strength	x	x		x	x		
Tensile strength	x						
Concentricity		x	x	x	x	x	

A reduced overview for the design of high-frequency strands (HF strands) is given in Table 2.8.

Table 2.8 Design matrix for high frequency litz wire development according to [Ele-15a]

	Design parameters				
Litz wire characteristics	Wire gauge	Number of conductors	Number of strands	Twist pitch length	Twist pitch direction
R_{DC}	x	x			
R_{AC}	x		x	x	
Outer diameter	x	x	x	x	x
Mech. flexibility	x	x	x	x	

Alternative conductor geometries

Besides the combination of different strands to a combined conductor, the individual geometry of the conductor cross section is another degree of freedom. The greatest benefit of using different conductor geometries lies in the fill factor, the ratio of conductor material to the available winding space. According to [Bra-11] and [Her-08], increasing the fill factor by 10 % leads, on average, to a one percent increase in motor efficiency. The manufacturing of different conductor profiles requires different manufacturing processes and/or downstream processes. For manufacturing, usually the bare conductor has to be rolled, as opposed to the otherwise common wire drawing. Alternatively it is also possible to roll the conductor into the desired shape after the circular conductor has been drawn to the correct gauge. The challenge in the manufacturing of profile conductors is the execution with the smallest possible edge radius for rectangular profiles, and the application of the insulation on the edges. These geometric properties lead to a more difficult handling of profile conductors, which is explained in more detail in Chapter 3.

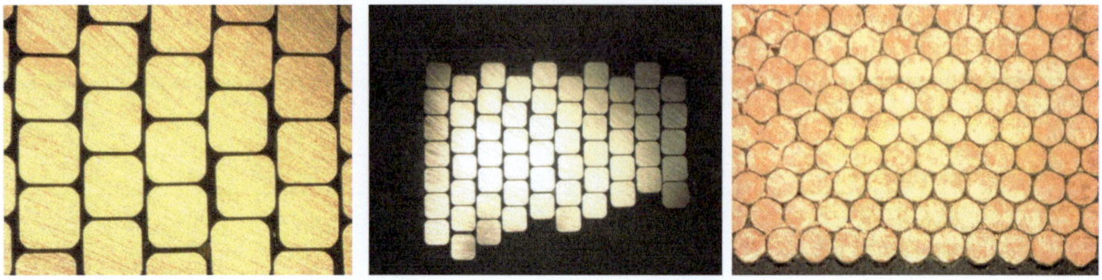

Figure 2.12 Cross section pictures for rectangular conductors and pressed circular wires

The most commonly used geometry is the circular conductor. This is due to the simple processing as compared to non-circular cross sections. Profile wires are only used for special applications. They enable a higher fill factor but processing becomes much more difficult due to the manufacturing properties.

2.2 Process chain for conductor manufacturing using the example of enamelled copper wire

Figure 2.13 Close up image of a wire drawing unit

The focus of this section is on the process chain for the raw material extraction of copper and the following processing to bare copper wire. The central process steps are described in terms of parameters, machine technology and possible process errors. The resulting physical properties of the bare wire are discussed. The manufacturing includes processes of forming and heat processing of the bare copper.

Accordingly, this section is directed at readers who are interested in the manufacturing of magnet wires. Since the magnet wire, as a central semi-finished product in winding technology, has an influence on the winding process, understanding the origins of the semi-finished product's properties is important for a suitable process design.

2.2.1 Process chain overview

The process chain for the production of magnet wire begins with the raw material extraction of copper. The source may be an ore mine, or material gathered from the recycling cycle. For a better understanding of the differences in the material quality, the raw material extraction from different sources is described further on. Depending on the region, other copper gathering procedures are possible. Figure 2.14 explains the different steps of typical copper preparation. Usually, natural copper ore is extracted from a chemical compound with sulphur as copper sulphide (Cu_2S), also referred to as copper glance or chalcopyrite ($CuFeS_2$), also referred to as copper pyrites.

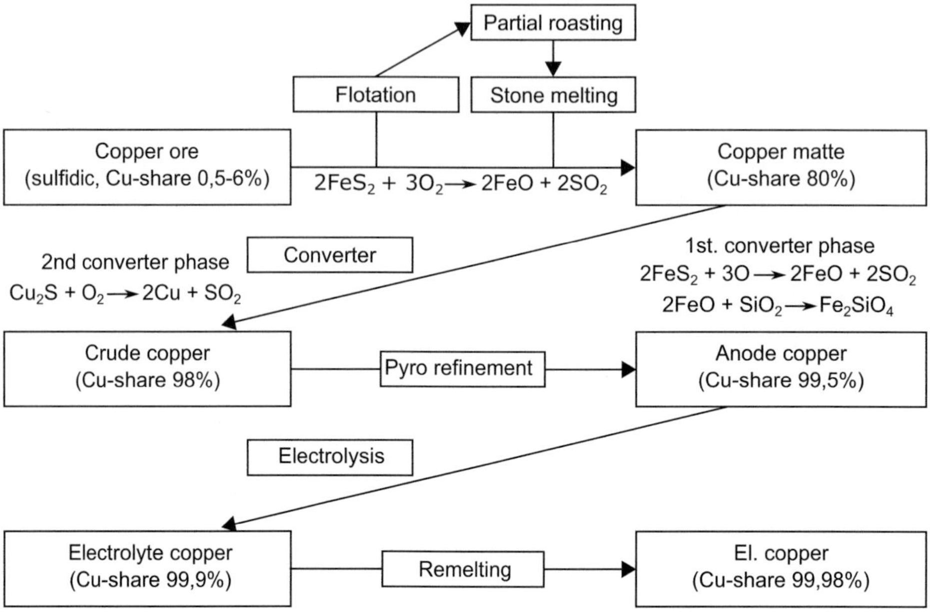

Figure 2.14 Process chain for copper purification according to [Ing-15]

At first, the raw ore only has a copper concentration of 2 %. Using the physical-chemical separation process of flotation, the accompanying rock is separated and the copper concentration increases to up to 25 %. Subsequently, the ore is roasted, where, under the addition of oxygen, sulphur dioxide and iron oxide are created in two converting phases, meaning two chemical process steps. After oxidation, the iron and the sulphur are bound in a swimming slag, and subsequently extracted by pouring. The remaining raw copper now has a concentration of around 98 %. For further enhancement the raw copper is cleaned in an electrolysis bath in which the (electrolytic) copper settles on the anode plate. Pure copper's first geometric shape is therefore always a plate. The transformation of copper shapes is shown in Figure 2.15. Subsequently, the copper plates are turned into bat material by re-melting. Due to the following rolling the bar material is reduced to a gauge of 7 to 8 mm.

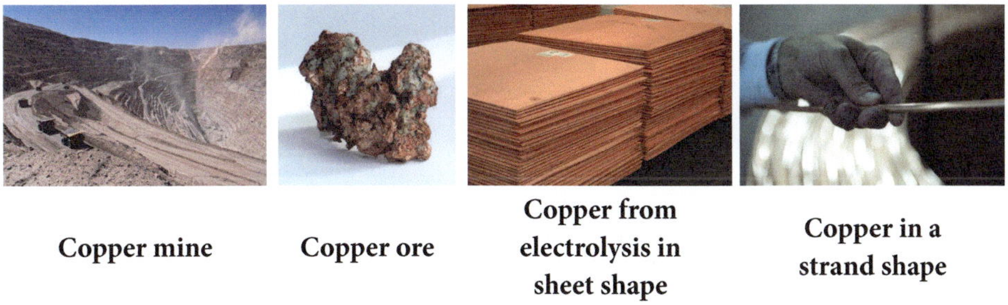

Figure 2.15 Different shapes of copper material: from the ore to the copper sheets (Pictures: M. Hagmann, Astro.GoBlack.de, Essex Germany GMBH)

The bar material mentioned above is usually the original shape for a typical wire manufacturing process chain. Generally, the manufacturing steps can be classified in the following processes; these are displayed in a wire drawing and enamelling machine in Figure 2.16:

- forming the wire to the desired gauge
- soft annealing of the wire
- cleaning the wire
- applying multiple insulation layers
- applying a lubricant or alternative functional layers: self-bonding enamel
- rolling up the wire on a supply spool.

The individual process steps and the respective machinery are covered in more detail in the following sections. The final roll-up of the wire onto a supply coil usually happens via a double spooler, where a supply spool is wound. Once the supply spool has reached its capacity, an automatic change to the reserve spool takes place. For the spool change the wire is stretched between the two rolls and tears, or for thicker wires exceeding 1.0 mm, additional tools are used. Subsequently, the free wire end is caught via a hook mechanism and clamped. The wire handling during the spool change takes place with catch hooks or catch springs and does not require an interruption of the manufacturing process.

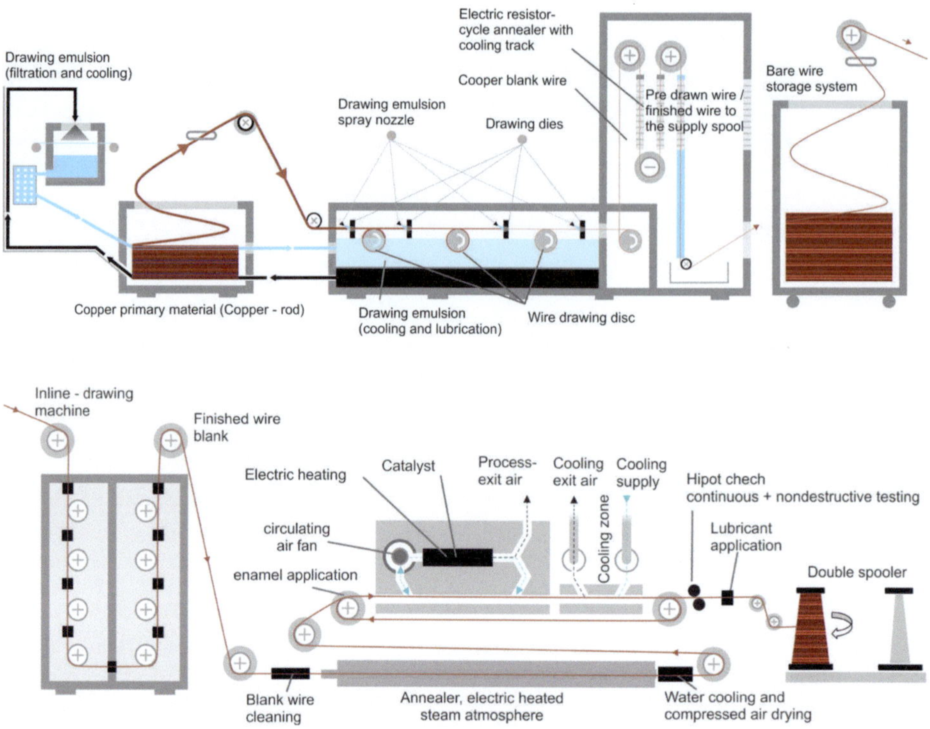

Figure 2.16 Process chain for magnet wire production

2.2.2 Drawing processes

The aim of the drawing process is to achieve a permanent change of the wire's gauge by exceeding the yield limit $R_{p0,2}$. The load required is reached by a continuous wire feed and the use of drawing dies. A drawing die consists of a circular metal disk as a base material with a diamond in the middle. The diamond has a hole, which approximately equals the final wire gauge. The hole itself contains an inlet cone, followed by a cylindrically shaped section and an outlet cone. For gauges up to around 0.15 mm, a natural diamond is used due to the better surface of the workpiece. Regardless of the wire gauge, this can also be found in every final drawing phase. Industrial diamonds are used for larger gauges exceeding 0.15 mm. Due to the spring-back or elastic resilience of the wire, it has a slightly larger gauge than the drawing die after the last drawing phase. This has to be considered in the design of the drawing phases. The structure of different drawing phases is illustrated in Figure 2.17. Typically, multiple drawing phases are used to ensure a gradual adjustment of the wire gauge.

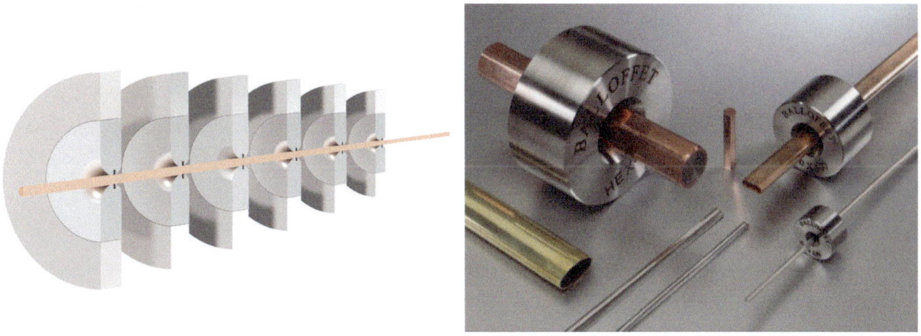

Figure 2.17 Structure of several drawing steps (right: BALLOFFET GmbH)

For fine wires, with gauges between 0.02 mm and 0.1 mm, 15 drawing phases are used. For wire gauges exceeding 0.1 mm, usually 21 drawing phases are used. To reduce the required space, they are arranged in multiple parallel phases on drawing cones. The drawing dies are located between the drawing cones. For decreasing wire gauges, the wire guidance takes place on increasing radii on the drawing cone. The basic structure, either two cones with 14 drawing phases each for thinner wires, or four cones with 10 drawing phases each for thicker wires can be used. A structure like this is shown in Figure 2.18. A serial arrangement of the drawing dies is also possible and can be used for gauges from 0.2 mm (Figure 2.17, left). The drawing cones themselves contain a grooved geometry, with the help of which the wire guidance is improved.

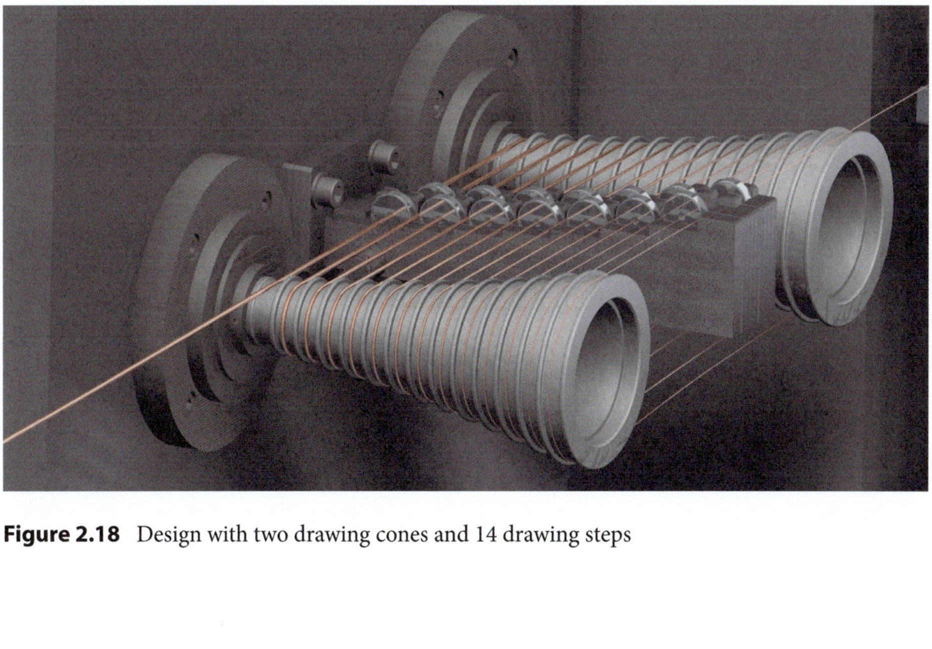

Figure 2.18 Design with two drawing cones and 14 drawing steps

Alternatively, the use of multiple drawing dies leads to a better wire guidance (Figure 2.18). Besides drawing cones, drawing disks can be used. These enable a vertical rather than horizontal arrangement. Different structural variants with drawing cones and drawing disks are shown in Figure 2.19. The drawing unit, the wire drive, usually consists of a mechanical transmission with a belt drive. The drawing dies have a gearing of speed and wire gauge of 8 % between, before, and after the drawing dies. The reduction of the wire gauge leads to a greater wire length, which has to be compensated by a higher feeding speed. Analogously, loose wires occur in front of the drawing die. On the drawing cone the circumference is only partly used to feed the wire. The gearing that arises from the drawing cones or drawing disks is 6 %. This results in a slip of 2 %, which can be used as a reserve. The slip itself is necessary in order to provide the loose wire length for the forming process and the compression when entering the drawing disk. The pairing of the gearings depends on the wire gauge, so that 10 % and 12 % also work as machine or wire elongation.

Design with two drawing cones **Design with drawing disks**

Figure 2.19 Comparison of different drawing setups

For thicker wires from 2 to 2.5 mm, gearings from 34 to 36 % can be used. Due to the forming with a high degree of deformation, large stress zones occur in the wire material. These necessitate the use of auxiliary materials, so-called drawing oils. The drawing oil ensures cooling of the wire, cleans it, and lubricates the drawing dies as well as the handling components (cones or disks) at the same time. The stresses on the wire during the drawing are immense, which is why it must be annealed after reaching the final wire gauge in order to reduce internal stresses. The drawing process depends on the wire gauge itself, the speed, and the required tolerances for gauges and elongation at break. Usually, only circular wire profiles are manufactured using the drawing process. Exceptions include complex geometries that are too filigree for the rolling process. One example is the hexagonal shape for a wire profile. The requirements for this forming process are much higher as those compared to circular wire. Therefore, only small wire gauges can be manufactured. In addition to the description of the drawing process, a complete setup of a wire drawing machine is shown in Figure 2.20.

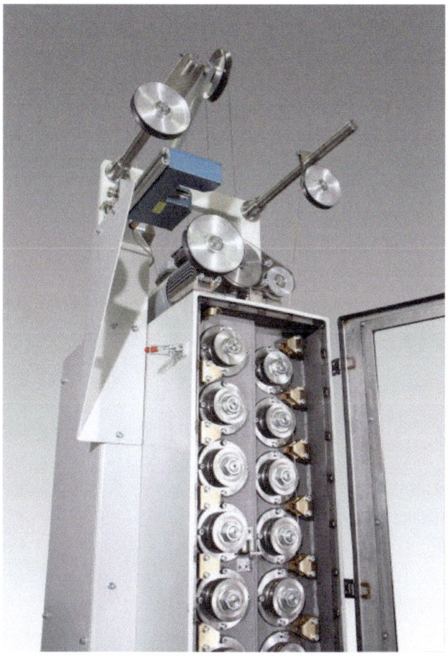

Figure 2.20 Complete setup of a wire drawing unit

The inline drawing process is a flexible process to manufacture desired wire gauges for different original gauges and creates a high-quality material surface with low surface roughness at very high process speeds.

2.2.3 Rolling process

In much the same way as the drawing process, the aim of the rolling process is to permanently change the wire gauge or height by pressing the wire through two pivoted rolls. The starting product for this process is also a circular wire. Usually, a longitudinal rolling process is used for the wire; this involves feeding the wire through the rolling gap perpendicular to the rolling axis without turning on its own axis.

The upper and lower roll turn in opposite directions. A profiling of the rolls creates a defined geometry. However, the rolling process always changes the wire profile. As opposed to drawing, where the profile is defined by the drawing dies, the use of the rolling process may lead to a deviation from the desired contour. Accordingly, the surface of the wire is skimmed to ensure the shape accuracy, generating a chip as by-product. Due to the special role of the rolling process, which is almost exclusively used for profile wires, it is not covered in more detail at this point. The structure of a pair of rolls for the manufacturing of a rectangular profile wire is shown in Figure 2.21.

Figure 2.21 Rolling pair for profile wire production (Picture: Essex Germany GMBH)

The rolling process is the preferred process for the manufacturing of profile wires. Vertical machines are used for the most effective manufacturing of profile wire.

2.2.4 Continuous annealing process

Forming the wire in its profile or gauge leads to strain hardening of the wire and a lower elongation at break, as illustrated in Table 2.9. To determine the values provided, the wire was stressed up to the given amount of its tensile strength.

Table 2.9 Resulting wire characteristics after wire drawing (CuMg0.7) [DKI-12]

Material state	Yield strength $R_{p0,2}$ in N/mm²	Tensile strength R_m in N/mm²	Elongation at break A in %
Annealed softly	80	270	50
Hard drawn (50 %)	410	450	5
Hard drawn (50 %)	510	550	4
Hard drawn (50 %)	640	695	2

Processing the wire without annealing would lead to malfunction during the winding process because its deformability is no longer given. Accordingly, the hardened material structure must be annealed to relieve the induced stresses. Therefore, the material must be heated up to a temperature above the recrystallisation limit in an annealing process. For wire production this is done with a continuous annealing process. For a proper annealing process, it is important that the temperature must be distributed down into the core of the wire through heat diffusion.

Depending on the wire thickness and the desired elongation at break, the temperature must be held for a longer or shorter period of time. The longer the temperature is held, the softer the wire will become. Certain temperature profiles are run for precise warming. Typically, two heating sections are used for the inlet and outlet temperature, while the pass-through generally maintains a temperature between 500 °C and 650 °C. This is based on the recrystallisation temperature of copper, which lies between 250 °C and 500 °C depending on the alloy. However, stress relieving begins between 150 °C and 200 °C. The annealing temperature also depends on the original hardness of the material and its composition, as illustrated in Figure 2.22 using the resulting hardness of different annealing temperatures as an example.

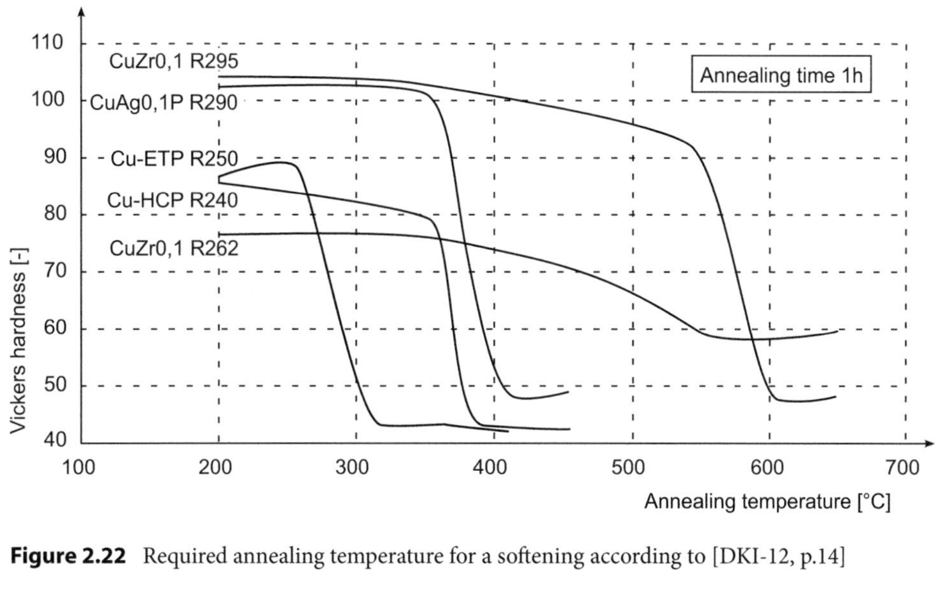

Figure 2.22 Required annealing temperature for a softening according to [DKI-12, p.14]

The length of the annealing section mainly depends on the wire gauge and can range from 1 m for thin wires, and up to 16 m for thick wires with gauges above 1 mm. The annealing section is brought to the desired temperature via an electrically heated tube. The annealing is followed by a cooling section using a shielding gas atmosphere to prevent the oxidation of the wire. Thin wires up to 0.1 mm wire gauge are cooled using a continuous media flow at room temperature. All gauges above 0.1 mm are water cooled. The settings for the desired elongation at break of the wire are adjusted by experiments during the commissioning of the machine. They are determined by tensile tests after the annealing. This is achieved by varying the temperature and the annealing duration via the conveyor speed.

However, due to the continuous process, the speed is prescribed by the slowest process, which is the following enamelling process. Accordingly, only the annealing temperature can be adjusted. The balance between the holding time and the temperature must be determined exactly (Figure 2.23, left). Annealing at lower temperatures also leads to long-term reduction of stresses as is shown in Figure 2.23 (right). At 200 °C the elongation at break can be increased from 3 to 20 %. An effective stress relieving temperature for this material (CuNiP), however, lies between 380 and 420 °C.

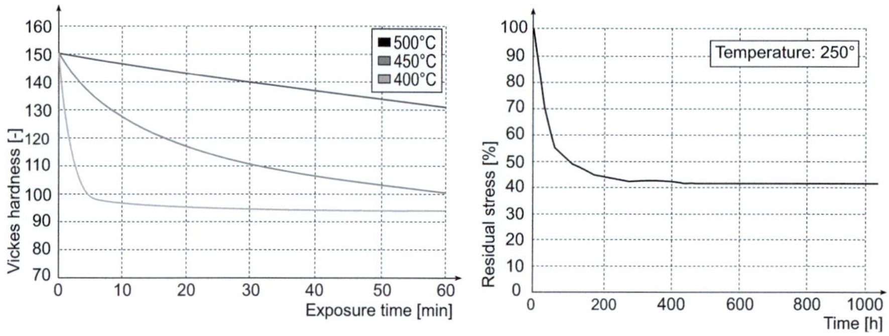

Figure 2.23 Influence of temperature and time on the annealing result (Left: CuFe2P, Right: CuNiP) according to [DKI-12, pp. 17 and 23]

In terms of the process chain, the advantage of annealing is that it simultaneously burns any drawing oil residue, which has a cleaning effect. The challenge of this process is the complete annealing at the shortest exposure time of the wire.

Annealing enables later deformability and leads to an initial cleaning of the wire for a flawless enamelling. Also, it creates the desired physical properties (strength and elongation at break) for the handling in the winding process and electrical resistance.

2.2.5 Cleaning processes

Cleaning the wire is necessary in order to ensure an even and functional enamel application. When foreign particles are coated, they can cause flaking of the enamel during processing of the wire, or due to ageing. This significantly increases the chances of a resulting component failure. Detecting unwanted particles during processing or on the product is impossible. However, the wire gauge and high-voltage strength can be determined as auxiliary quantities inline of the wire manufacturing, or the winding process. Foreign particles may be, for example, drawing oil, dust from the drawing process, surface contaminations of the original material, or dust from the production environment. For thin wires up to 0.2 mm wire gauge, cleaning by the continuous annealing is sufficient. For thicker wires, the cleaning is done with a water bath in a tank. Additionally, a sponge is used for cleaning inside the water bath in order to remove the remaining particles through friction.

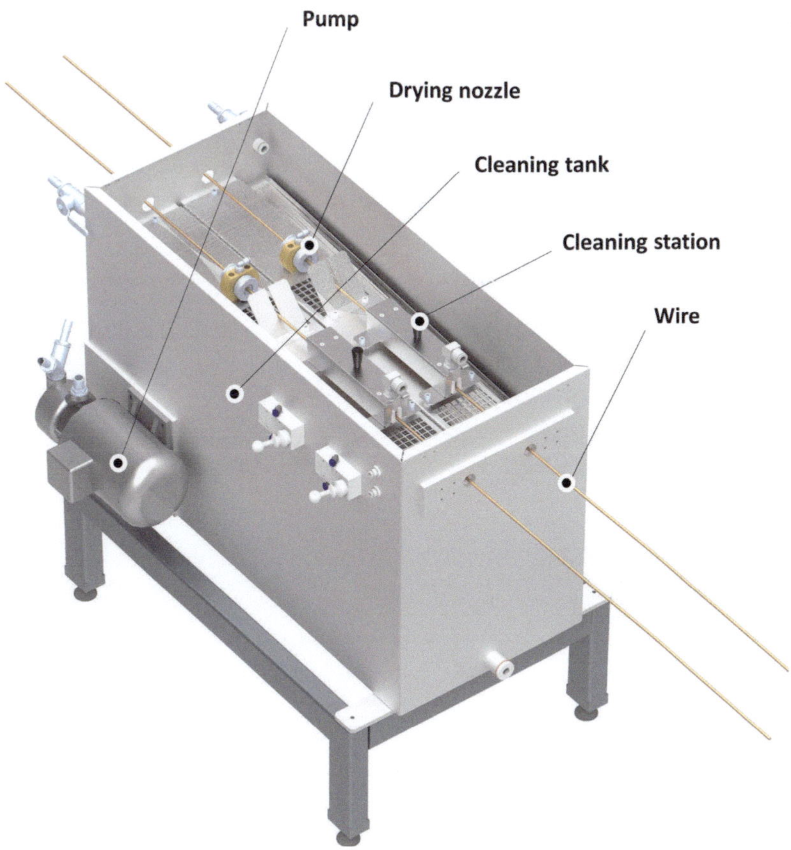

Figure 2.24 Structure of the cleaning unit

The sponge is continuously rinsed with water to ensure that the particles are removed. This process can be used for various conductor materials. A schematic illustration of a cleaning station can be found in Figure 2.24. Aluminium wires are an exception; they are cleaned in an ultrasonic bath. This is because the surface of aluminium wires risks being damaged by the sponge, which may lead to the removal of the protective oxide layer. Accordingly, contact-free processes are preferred.

The cleaning process is a key requirement for the optimal application of the enamel on the wire and minimises the influences of foreign particles in the following processes. Especially for thermally and mechanically highly demanding winding goods, the cleaning of the wire before the enamelling process is a necessity.

2.3 Conductor insulation

Figure 2.25 Nozzle varnishing of the wire

The process chain in the previous sections described the manufacturing of the pure bare wire. In Section 2.3, the insulation layer that is to be applied is discussed in terms of its product and manufacturing properties. It ranges from insulation properties to the introduction of suitable testing procedures for different insulation materials and different enamelling processes. The insulation materials are only briefly described, since they only differ slightly in terms of their application. Special focus is placed on the introduction of typical error patters from the enamelling process.

Accordingly, this section is directed at readers, who are interested in the manufacturing of magnet wire. Due to the fact that the magnet wire as a key semi-finished product has an influence on the winding process, understanding the origins of this semi-finished product is important for a suitable process design. A coil's electrical properties are defined to a large degree by the insulation and its application. Therefore, this section is also relevant for readers with a development or quality control background.

2.3.1 Definition of insulation properties and testing procedures

Introduction

The electrical insulation of the wire is a key property for the use of enamelled copper wires in coils. The execution of the insulation complies with the application's requirements, such as the operating voltage, the operating current or the winding schematic. Besides the electrical requirements, however, there are also mechanical, chemical and thermal requirements for the insulation. Basic terms and properties of insulation technology for magnet wires are described thoroughly in [Syn-15].

A central parameter for the characterisation of insulation properties is the breakdown voltage. It defines the insulation strength of materials. If an insulation material is contacted with two electrodes and a rising voltage is applied, a breakdown occurs when the insulation strength is reached, and the insulator becomes electrically conductive. For enamel insulated copper wires, the breakdown voltage depends on the thickness of the enamel layer, or rather the equal application, the centricity. The surface quality of the bare and the annealing grade of the enamel layer are additional influencing quantities. The degree of annealing is a measure for the correct evaporation of the previously applied enamel. The electrical and mechanical insulation properties for different enamel types are standardised in IEC 60317. The respective testing procedures are described in IEC 60851. The insulation thickness and the respective electrical effectiveness are classified in levels. For each level results a specific ratio of the copper cross sectional area to total cross sectional area. This is shown in Figure 2.26. The insulation levels describe a minimum and maximum value for the increase of the layer thickness. Level 1 is the thinnest, and level 3 is the thickest insulation layer. The layer thicknesses merge into each other. Level 3 is usually used for applications with high reliabilities, such as wind turbines. The introduction of two further insulation levels, 4 and 5, is expected for the future due to the trend towards higher voltages for different applications. [Syn-15]

Dimensions and electrical properties are standardised in IEC 60317. This includes the minimal breakdown voltage, the DC resistance and the number of allowed imperfections of the insulation as shown in Table 2.2. Besides the product properties, the processing properties are also important for the choice of insulation. These are relevant for the desired contacting processes (Section 3.4.2), or for the application of a secondary insulation of the coil with a resin (Section 3.4.6).

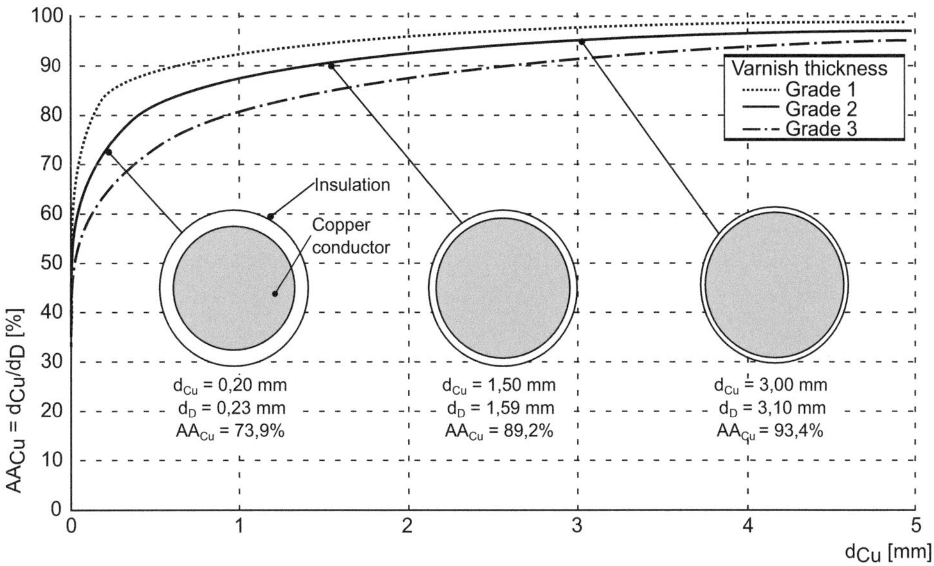

Figure 2.26 Wire gauge depending on insulation layer properties according to [Dob-11, p. 28]

Testing procedures for magnet wires

To measure the breakdown voltage, the stripped wire ends of a wire probe are clamped into a test bench. The wire is lead over a roll and stressed by tensioning it. In the next step, a high voltage is applied to the wire and increased until the insulation malfunctions and an electrical breakdown occurs. The breakdown takes place between the roll and the wire and can be detected on the wire ends by a sudden voltage drop. The breakdown voltage is standardised according to IEC 60851-5/4/2012-04, and is a measure for the insulation strength.

Another test is the tangent-δ test according to IEC 60851. The term tangent-δ describes the dielectric dissipation factor, which only applies for dielectric media, such as the insulation. As opposed to theoretically ideal insulations, real insulations show small dielectric losses. As an analogy, the charging of capacitor plates can be used, where losses occur during the charging cycle. The losses are caused by dipole friction on a molecular level. Vibrations caused by the alternating voltage produce heat due to friction. The losses depend on both the degree of annealing and the chemical structure of the insulation layer. The testing takes place with a contacted and drawn wire. Therefore, the test arrangement consists of a drawing machine, which is equipped either with two drawing disks, aligning behind each other, or drawing cones. The arrangement of the drawing disks is referred to as a tandem arrangement. Wire elongation is achieved by applying different conveyor speeds to the drawing disks. Subsequently, the molecular structure of the insulation is stimulated by increasing the temperature until the dielectric losses increase. Both quantities are continuously recorded. The tangent-δ breaking point in Figure 2.27 marks the point at which the losses increase (IEC 60851-5/Appendix A/2012-4).

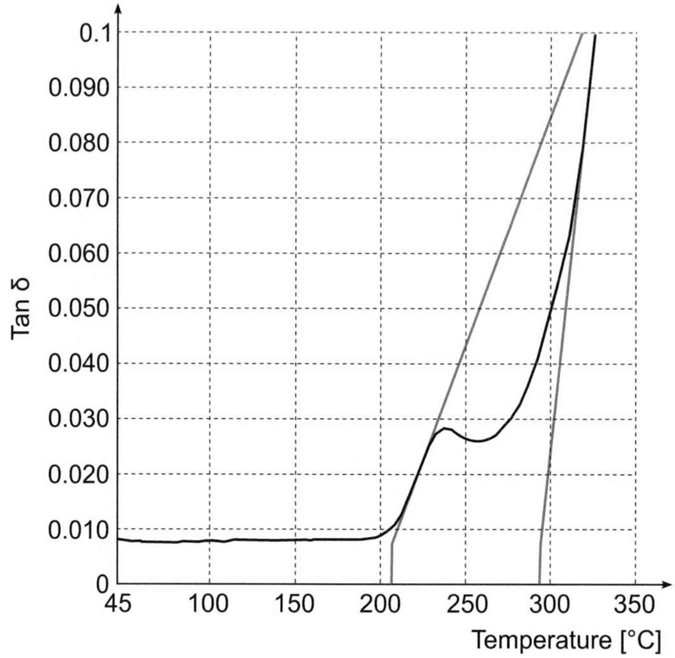

Figure 2.27 Characteristic tangent-δ measurement

This point describes the thermal malfunction of the insulation and provides a characteristic regarding the degree of annealing. As a rule, a higher degree of annealing in the wire manufacturing causes the breaking point to shift towards a higher temperature. The evaluation of the measurement can be done according to three criteria [Syn-15]:

- tangent-δ value at a fifth of the rated voltage
- increase of the tangent-δ value from the beginning of the test until malfunction of the insulation
- characterisation of the tangent-δ progress.

Another test for the insulation properties is the testing of the re-melting temperature: Two wires are crossed and connected at their ends. Subsequently, a stamp pushes down onto the crossed arrangement. The temperature of the wire including its insulation is increased until the wire re-melts and the insulation malfunctions.

A common test for magnet wires is the pinhole test according to IEC 60851-5/7/2012-4, also referred to as imperfection count test. Its goal is to determine the number of imperfections in the wire insulation. This test can be performed with a device for checking defects. Alternatively, it can be determined with a water bath in a batch process, which is a non-continuous process. The wire is then inserted into the water bath and contacted at the wire ends. Subsequently, the wire probes are stored at dried at 125 °C in a ventilated cabinet for 10 minutes. A salt bath with 2 g/l NaCl and a phenolphthalein solution in the water follows. In the bath, a measurement at a voltage of 12 V is carried out for at least one minute. If there is an imperfection in the insulation, it is revealed by a violet stain in the solution. [Syn-15]

Another type of test is the wire helix test according to IEC 60851-3/5/2014-7. The term wire helix describes the formation of a wire loop. This is wound around a mandrel, which is where a helix forms, and then loaded up to a 60 % elongation of the outer fiber. Subsequently, the insulation of the wire helix is inspected for cracks with a microscope. Finally, the elongation at break is determined with the tensile test. The enamel's adhesion on the bare wire can be determined at the breaking point. Poor adhesion enables the enamel layer "to be pulled off of the bare wire like a straw" [Syn-15].

There are further procedures for testing, however, they are not individually covered here. This includes the testing procedure for the determination of the temperature index of magnet wires according to IEC 60172/2010-11, and the determination of the high-voltage imperfection count according to IEC60851-5/5/29012-04.

For testing insulation requirements, specific testing procedures are available. Most of them are based on random samples, or are out-of-process tests. The number of allowed imperfections is standardised, and due to process monitoring and process control it is usually even undercut.

2.3.2 Introduction to insulation materials

The electrical insulation of a wire has an influence on different product properties, such as the area of application, durability, costs, or coil design as well as different manufacturing properties, such as rejects, processing or suitable winding processes [Bec-13]. The insulation materials contain different amounts of solids. These indicate the volume ratio of synthetic resin in the enamel. Enamel with 40 % solid content has 60 % solvent and 40 % insulation material. For thin wires a solid content of 20 % is common. For thicker wires the solid content is increased to 45 %. [Syn-15] Generally, the following insulation materials can be differentiated:

- solderable enamels, such as polyurethane
- heat resistant enamels, such as polyester
- solvent resistant enamels, such as polyester amide imide
- self-bonding enamels, such as polyester amide
- nylon
- textile insulation for wires, such as electric glass fibres, alkali-free and enamel-impregnated
- tape and foil insulation, such as mica/polyester film as tape with epoxy resin.

Typical insulation materials and their properties are described in Table 2.10. The temperature index indicates the insulation's maximal operating temperature. The heat shock temperature describes a possible, temporally limited excess of the temperature index. This is determined with a wire helix wound on a testing mandrel in a heat cabinet. Similar to the adhesion test, the insulation is inspected for possible cracks using a microscope after the heat test. The softening temperature indicates the enamel's melting temperature. Table 2.11 lists a large variety of existing insulations systems and their applications. A current trend is the integration of additional functionalities into the insulation material, for example, the integration of lubricant into the enamel in order to avoid a separate coating.

Table 2.10 Comparison of different insulation materials [Bec-13]

Material	Temperature Index	Heat shock	Re-softening temperature	Tinning	Chem. Durability
Polyvinyl acetate	120 °C	155 °C	> 215 °C	No	Excellent
Polyurethane	180 °C	200 °C	> 250 °C	Yes	Good
T. Poly.(imide) + Polyamide-imide	200 °C	220 °C	> 340 °C	No	Good
Polyamide-imide	220 °C	240 °C	> 400 °C	No	Excellent
Polyimide	240 °C	260 °C	> 500 °C	No	Excellent

There are a large number of different insulation manufacturers, and this section will only cover a few insulation materials and their properties.

Table 2.11 Application of insulation materials [Von-12]

Application	Varnishes round wires	Varnishes flat wires	Woven wires	Band- and foil insu-lated wires	Varnishes litz wires
Transformers and switchboard plants	x	x			x
Electric tools	x				
LV-motors (industry)	x		x	x	
LV-motors (automotive)	x	x			
Electric industry	x				x
Traction drives	x	x	x	x	
Generators for wind turbines			x	x	x
High voltage rotational machines			x	x	x
Large generators			x		
Repair industry		x	x	x	

Typical wire manufacturers are Essex, Schwering und Hasse, Heerman Wickeldraht or Elektrosiola, or von Roll. Typical insulation manufacturers for magnet wire are DuPont or Altana.

There are numerous insulation materials, although the majority is based on resin. The application determines the required insulation class, which results in different thicknesses and materials of the insulation layer. The chosen insulation, however, only plays a minor role in wire processing, except for lubrication.

2.3.3 Enamelling processes for insulation application

The enamelling process is either performed with a felt, which applies the enamel, or a nozzle. From a supply tank the enamel is fed to the felt or the nozzle with a gear pump. At first, the felt is soaked until saturation (Figure 2.28). Then the enamel is applied to the wire, keeping in mind that the volume fed by the pump corresponds to the enamel applied to the wire. A hole in the nozzle determines the flow quantity. The actual enamelling takes place in front of the nozzle. For larger applications the nozzle's diameter is increased accordingly.

Nozzle varnishing

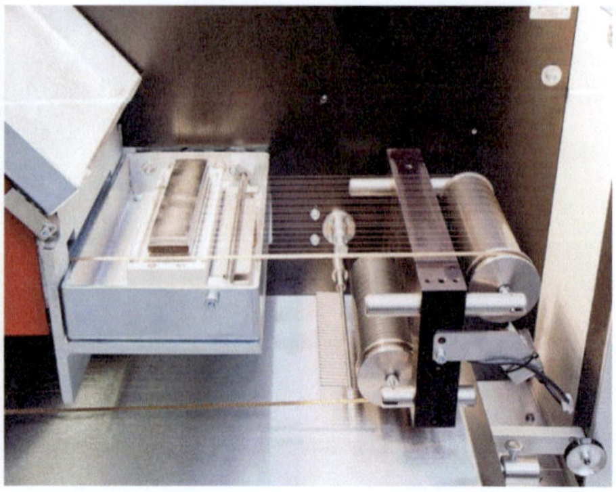

Felt-varnish application

Figure 2.28 Process pictures. Scan QR code to view demonstration video.

In terms of machine design there are two possible types of enamelling machines. A differentiation in design is made due to the diameter-dependent bending of the wire, which can impede the wire guidance. A horizontal design is used for thin wires with gauges between around 0.001 and 1.2 mm. A vertical design is used for wire gauges from 0.5 to 5 mm. The machine components are similar for both types. The wire supply to the machine, generally referred to as wire feeding, is done from a spool (thin wire) or a drum (thick wire). Before enamelling, the wire is drawn using an inline drawing unit in order to ensure a clean and high-quality surface. Subsequently, another cleaning, cooling and lubricating unit is necessary due to the drawing emulsion. In the annealing unit the wire is again annealed, and then cooled. Subsequently, the actual application and annealing of the enamel takes place. This is achieved by applying the enamel in very thin layers of around 0.5 to 1 µm. The number of layers depends on the wire gauge. Generally, a nozzle can be used to apply the enamel for wire gauges from 0.15 mm. It also enables the application of more enamel than the felt:

- 12–24 layers for thin wire gauges 0.012 to 0.03 mm
- 15–30 layers for medium wire gauges 0.04 to 0.15 mm
- 12–24 layers for thick wire gauges 0.15 to 5 mm

The enamel application is followed by the drying and cooling phase for the wire. The insulation itself is only mechanically resilient after it has cooled down. Due to the wire guiding via pulleys, the cool down must take place before the wire can be processed further. The coating process is repeated up to the desired enamel thickness. Generally, the enamelling process is similar for different wires; however, it varies in different temperature profiles for the drying of the enamel depending on the insulation material. The previously described technique for enamel application can be used for the primer, which is a base enamel for better adhesion, as well as for the overcoat, which is the actual insulation enamel, and for the application of self-bonding enamel for bonding coils. The wire gauge can be determined using laser metrology according to the shadow principle in the machine. The enamelling results can be characterised based on different test procedures as described in the previous Section 2.3.1. A common measure for characterisation of the insulation quality of the wire is the continuous pinhole test. As opposed to the previously described batch process with a water bath, for the inline test a set of pulleys or brushes is used to contact the wire with a voltage in order to identify imperfections. In the case of too many faulty insulation sections, the batch is rejected and the process must be adjusted.

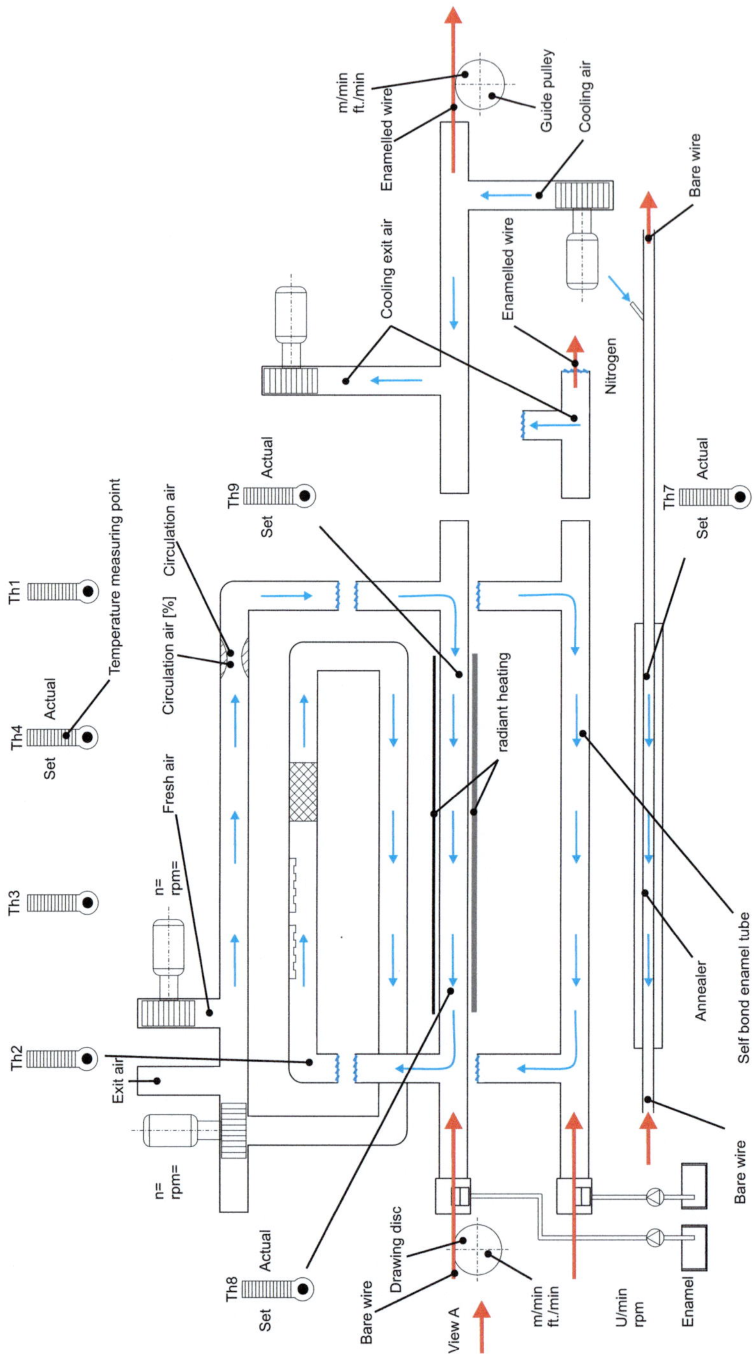

Figure 2.29 Structure and media flows for a wire varnishing machine

All of the key process parameters are recorded in the machine or the central control of the machine, in order to be able to analyse batch fluctuations. Key parameters of the enamelling are the volumetric flow rate of the geared pump, and the nozzle diameter for the desired amount of enamel. Temperature, wire speed and the volumetric flow rate of the drying air, which is responsible for the hardening of the wire, are also crucial. Several typical process errors in wire enamelling and common respective causes are illustrated in Table 2.12 and Figure 2.30.

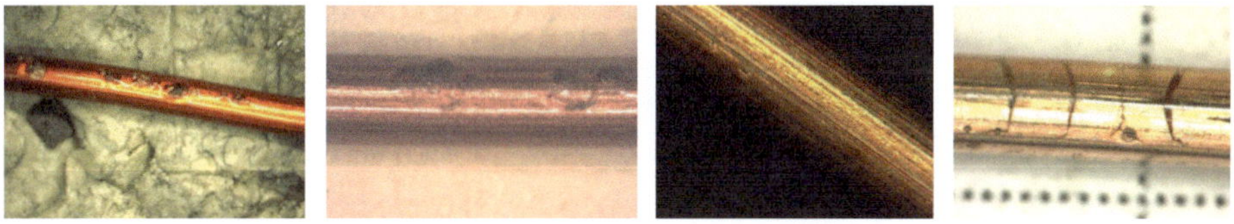

Contamination of the bare wire | **Blistering** | **Roping** | **Cracking**

Figure 2.30 Typical error patterns for the production of varnished magnet wire

A typical error pattern is the development of bubbles on the surface of the enamelled wire. They are often caused by inclusions in the enamel layer, such as solvent or humidity, which can come from the cooling section of the annealing unit, for example. The inclusion of foreign materials leads to an evaporation of the material during annealing, creating bubbles on the wire surface.

Table 2.12 Typical error patterns for varnished magnet wire

Error	Cause
Blister on the wire surface	– Contamination of the wire surface before varnishing
Non-adhesive varnish: Rough surface	– Process speed to high – Temperature too low – Worn out felt / wrong type of nozzle
Wrong elongation at break	– Too high or too low annealing temperature
Wrong friction characteristics: Problems with wire feeding	– Problems at the lubricant application
Faulty insulation: Partial-Discharge and dielectric strength	– Example: scratching of wire on rolls or machine elements may lead to varnish removal, or comparable causes to blister occurrence – Non-sufficient insulation drying

The enamelling of shaped wires is executed with profile nozzles and not with felt. The problematic area during enamelling is the edges of the profile wires, and the necessary even layer thickness, such as for hexagons.

The enamelling process strongly depends on the desired wire properties (gauge and temperature class, insulation class), and is monitored and recorded if possible. Gauge and insulation properties can be tested during the process. Further properties can only be determined after the enamelling process. The inlet wire properties, such as hardness and gauge, must be provided with consistent quality, otherwise fluctuating batches may result. Due to the modified finished wire properties this can lead to rejection in a worst case scenario.

2.4 Use of lubricants

Figure 2.31 Application tool for lubricants

The previous process chain described the manufacturing steps for the insulated copper wires. For its processing it is necessary to keep the friction between handling elements and the wire as low as possible, in order to prevent unwanted deformation. In this chapter the aspects of different lubricants and possible applications are discussed.

This section is directed at readers who are interested in the manufacturing of magnet wires. Since the wire, as a key semi-finished product in winding technology, has an influence on the winding process, understanding the origins of the semi-finished product's properties is important for a suitable process design.

2.4.1 Definition of lubricant properties

Processing magnet wires without lubricant additive would not be possible due to the friction properties of the insulation. This is why a thin layer of lubricant is applied after each enamelling process. It consists of paraffin or a dissolved form of bee wax. A 0.5 to 1 % benzene solution is used to dissolve these solid materials. The application to the wires takes place with a soaked thread. The amount actually applied is nominal. To 1 km of enamel wire with a gauge of 0.5 mm only 40 mg paraffin are applied. Accordingly, for 1 kg of wire this means 22 mg paraffin. A trend in the development of insulation systems is the integration of better lubricating properties into the insulation materials in order to be able to omit the addition of lubricants in the form of paraffin.

A lack of lubrication can lead to knots during the unwinding from supply spools due to slacks or holes. Due to higher friction, supply spools with parallel flanges cannot be used. Instead, bi-conical flanges are used which enable an over-head pull-off (Figure 2.32).

Figure 2.32 Wire feed from supply spool

In case of too little lubrication, the dullness or high coefficient of friction of the wire can lead to crossings on the supply spool. Due to this wedging, pulling the wire off radially can be very difficult. An axial pull-off is not possible at all. While processing, problems in the winding machine can occur when the exit angle is between 60° and 90°. Due to the friction between wire and wire guiding, a wire tear can occur as well. Generally, the wire tension will inevitably increase in the winding machine. This makes damage to the insulation in processing more likely, for example by shearing. At higher ambient temperatures too much lubrication on thin wire gauges leads to depositing of the material at the nozzles. This can lead to congestion and eventually to a wire tear.

In the following processes, problems of compatibility with the secondary insulation may occur, because the insulation resin cannot adhere and therefore causes imperfections. Similarly, the lubricant can lead to imperfections in combination with self-bonding enamel wires, because a mechanical connection is supposed to be achieved.

An accurately dosed lubricant is essential for processing magnet wires. Alternatively, top coats can be used, although the lubricating properties of the electrical insulation layer are specifically changed. Due to small margins, they do not have a higher market distribution.

2.4.2 Introduction to lubricant materials

The previously mentioned paraffin-benzene solution has an organic solvent concentration of 98 to 99.5 %. When using a wax, such as bee wax, the concentration can be 50 to 95 % or, alternatively, even completely solvent-free. Due to the necessity of collecting and burning the solvent, solvent-free solutions are more environmentally friendly. Primarily though, the customer requirements for an even application and thickness of the lubricant are paramount. The thinner the wire is, the less suitable are the currently available solvent-free systems.

The choice of lubricants for magnet wire is highly limited. This is due to the required durable and constant adhesion properties of the material. A creep behaviour cannot be tolerated in this context, because one must be able to process the wire even after a long period of storage.

2.4.3 Processes for lubricant application

A distinction in the application process is made between fluid and solid application. The fluid application is performed using wax which is dissolved in benzene and fed into the aforementioned felt with a pump. The disadvantage of the fluid application are the benzene vapours, which must be extracted, and entail special machine regulations. The advantage of this process is that it can be performed regardless of the wire gauge.

The solid application is performed with a wool thread soaked with lubricant, which entwines the continuously moving wire multiples times (Figure 2.33). If the wire is fed, the frictional heat liquefies the wax and applies the lubricant. For thinner wires, a wire tear also leads to a thread tear due to the entwining. The application is performed once, and averages between 5 and 80 mg/m^2 on the wire surface. The disadvantage of this process is a minimal wire gauge of 0.05 mm and the demand for wearing semi-finished products. The advantage is the omission of benzene vapour due to the missing liquefaction.

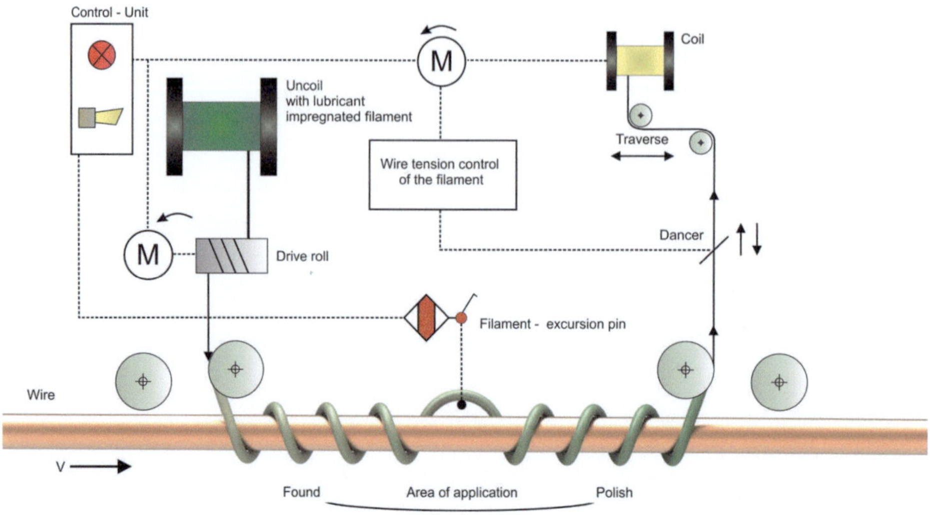

Figure 2.33 Setup for lubricant application

To check the lubricant application, there are testing devices to determine the coefficient of friction for samples. This is achieved by feeding the wire from one spool to another. Between the spools the wire is fed through a structure of horizontally arranged metal plates. The upper metal plate is loaded with a weight, and the resulting wire tension can be used to determine the friction properties.

The amount of lubricant itself can also be measured with a fine scale. As a current research aspect, the application by vapour or spraying of waxes can be stated.

The application of lubricant is necessary to ensure the manageability of (un-)winding, wire guiding in the machine, and in the winding process. The decisive factor being the amount applied and the resulting lubrication properties.

References

[Bec-13] Beckmöller S: Wickeldrähte - ein High-Tech-Produkt. Möglichkeiten und Grenzen. Fachworkshop FAPS TT, Nürnberg, 24.04.2013

[Bic-14] Bickel B, Hübner M, Franke J: Analyse des Optimierungspotenzials zur Erhöhung des Kupferfüllfaktors in elektrischen Maschinen. In: ant Journal 53 (2014) Nr. 2, pp. 16–21

[Br-11] Braun M: Motoren mit hohem Wirkungsgrad - Optimierung durch Berechnung und Fertigung. ZVEI (Ed.): 2. Fachtagung zu Produkten und Produktions technologien für die Herstellung von Elektromotoren, Generatoren und Transformatoren 2011, pp. 31–37

[Dem-11] Dempewolf K-H: Herausforderungen und Probleme bei der Fertigung von E-Maschinen für die Automobilindustrie. Contribution wbk Herbsttagung, 27.10.2011

[DKI-00] German Copper Institute: Kupfer in der Elektrotechnik - Kabel und Leitungen. URL: www.kupferinstitut.de (11.08.2015)

[DKI-12] German Copper Institute: Niedriglegierte Kupferwerkstoffe. URL: www.kupferinstitut.de (11.08.2015)

[Dob-11] Dobroschke A: Flexible Automatisierungslösungen für die Fertigung wickeltechnischer Produkte. ISBN 978-3-87525-317-7

[Ele-15] Elektroisola: Overview conductor materials. URL: http://www.elektrisola.com/de/leitermaterial.html (11.08.2015)

[Ele-15a] Elektroisola: Choice for litz wire parameters. URL: http://www.elektrisola.com/de/hf-litze/begriffe-grundlagen/auswahl-der-litzen-parameter.html (11.08.2015)

[Her-08] Herzog H-G: Effizienzpotentiale bei Elektromotoren - Status und Ausblick aus Sicht der Forschung. Kooperationsforum Energieeffiziente elektrische Antriebe 2008

[Ing-15] URL: https://www.ingenieurkurse.de/werkstofftechnik-2/werkstoffe-auf-eisenbasis/herstellung-von-metallen/weitere-metallische-werkstoffe/kupfer.html (11.08.2015)

[Ku-09] Kugler H: Umformtechnik: Umformen metallischer Konstruktionswerkstoffe. ISBN 978-3446406728. Hanser Verlag, München 2009

[Syn-15] Synflex: Dictionary. URL http://www.synflex.com/de/fachlexikon/ (11.08.2015)

[Tif-07] Ivers-Tiffée E, von München W: Werkstoffe der Elektrotechnik. ISBN 978-3-8351-0052-7. Teubner Verlag 2007

[Von-12] Von Roll: Wicklungsdrähte und Litzen. URL: http://www.vonroll.de/media/files/downloads/broschures/Wire_DE_20120516.pdf (12.08.2015)

Winding technology 3

The aim of this chapter is to provide the reader with knowledge about key winding processes and the respective machine technology. The specifically designed winding schematics lead to challenges in terms of production technology. Aids to design processes and machine technology especially for those particular challenges will be introduced. The winding process as part of a process chain is presented with regard to possible following processes and their parameters.

Figure 3.1 Machine elements for the production of coil winding products

3.1 Basics of winding technology

The aim of this section is to discuss the physical basics for this forming based assembly process. Based on this, guidelines for element design containing wires are presented. As a key unit in winding technology, the fill factor is introduced. Subsequently, different schematics for the winding design are explained, and their advantages and disadvantages are compared regarding operation and manufacturing.

This section provides assistance for the design and manufacturing of different coil schematics.

3.1.1 Physical basics

As described in the first chapter, winding is a process based on the forming of wire-shaped objects. Accordingly, many of the winding process properties involve the forming properties of the wire. This is especially relevant for the achievable packing density of the winding.

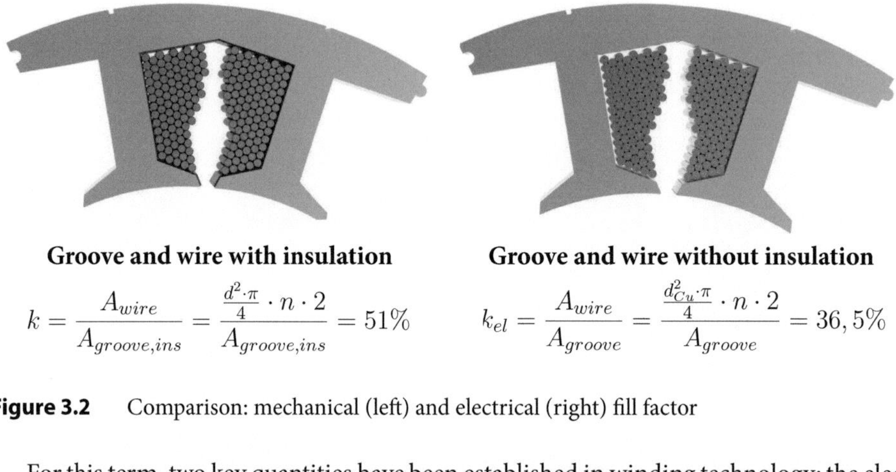

Groove and wire with insulation

$$k = \frac{A_{wire}}{A_{groove,ins}} = \frac{\frac{d^2 \cdot \pi}{4} \cdot n \cdot 2}{A_{groove,ins}} = 51\%$$

Groove and wire without insulation

$$k_{el} = \frac{A_{wire}}{A_{groove}} = \frac{\frac{d_{Cu}^2 \cdot \pi}{4} \cdot n \cdot 2}{A_{groove}} = 36,5\%$$

Figure 3.2 Comparison: mechanical (left) and electrical (right) fill factor

For this term, two key quantities have been established in winding technology: the electrical and mechanical fill factor. The ratio of the amount of insulated electrical conductor material, measured by the cross sectional area, to the available winding space is referred to as mechanical fill factor. This is determined by setting into relation the sum of the electrical conductors including their insulation with the available insulated winding space. A comparison of both quantities is shown in Figure 3.2. Below the illustration, a sample calculation is displayed for each fill factor arrangement. In the given example, the coil contains n = 38 windings. The bare wire's diameter is d_{Cu} = 0.5 mm. With the insulation, the diameter is increased to d = 0.548 mm. The available winding space is determined to A_{Groove} = 41 mm², defined by the full lamination stack. If the slot is insulated with an insulation thickness of 0.25 mm, the winding space is reduced to $A_{Groove\ ins}$ = 35 mm². The calculated fill factors were chosen as examples, but demonstrate a general problem. A direct winding of stators is difficult, because additional space must be left for the wire guide nozzle to place the wires in the groove. The term *direct winding* describes a winding that is manufactured at its place of function rather than being separately mounted on a further bobbin or winding tool. Accordingly, this space cannot be used for the winding itself.

Especially with round electrical conductors it can be derived that an organised structure, also referred to as layer structure, only leaves small spaces of the winding unused, leading to a higher fill factor and to better efficiency of the electrical component. As a round wire always includes spaces that are not electrically effective, the fill factor is always smaller than 1. Generally, wires with a thicker insulation layer for higher voltages provide the same or a higher mechanical fill factor, however, at the same number of ampere-turns, they also entail a higher resistance of the winding. This is due to larger coil dimensions with the same copper area per slot cross section. Because of larger wire diameters and the same number of turns, the winding height increases, which leads to an increased wire length on the coil. Measures to achieve high degrees of filling using a specific winding schematic designs are discussed in Section 3.1.2.

However, the basis for achieving a good fill factor is always the specific forming of the wire for the desired winding structure. The stresses on the wire and the resulting forming properties play a decisive role here. Typical stresses on the wire are tensile, compressive and bending loads in the winding process. The emerging stresses in the wire and the resulting changes in shape are illustrated in Figure 3.3.

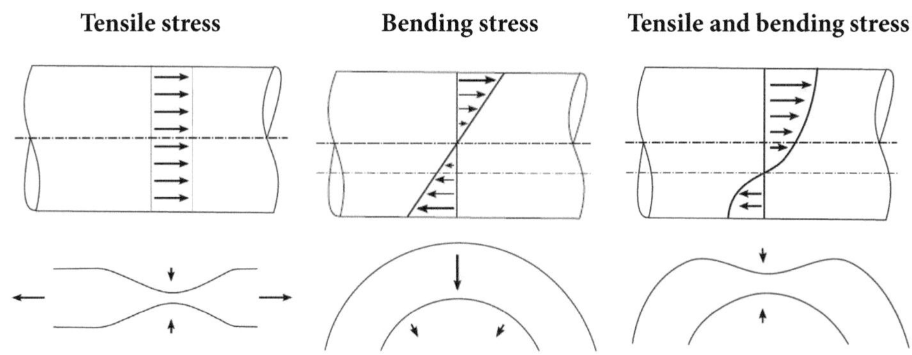

Figure 3.3 Mechanical stresses on the wire during the process

Tensile stresses occur during all winding processes. The causes are relative movements between the wire and the bobbin during winding. When the wire is set into motion, its mass, and the friction in the wire guiding elements of the machine, lead to minor wire tension. Compared to the inherent tension, it is significantly smaller and therefore not sufficient for the winding process. An additional wire base tension is therefore applied to ensure a defined positioning of the wire on the bobbin. If the additional wire base tension were not applied, the inherent tension, which originates from manufacturing and storing on supply spools, would lead to an undefined wire placement and bending contour. Besides the additional wire base tension, dynamic disturbances may result from the process as well. The winding machine must balance these out. The responsible machine component is the wire brake, which will be introduced in Section 3.2.1. The necessity of wire tension control lies in the prevention of a too large wire gauge tapering, which can lead to an increased winding resistance. Tensile stresses up to the wire's yield point $R_{p0,2}$, the elastic region, are tolerated during the winding process. Stresses beyond should generally be avoided, because the plastic deformation leads to a permanent wire gauge tapering and an increase in the winding resistance. In exceptional cases, for instance due to high demands to the process dynamics or hard to wind bobbins, wires with larger original diameters are used. Theseare then reduced to the desired wire diameter by deliberately exceeding the yield limit in the winding process. But the insulation's functional capabilities must be ensured and tested in advance. If the wire stresses during the winding process exceed the tensile strength of the copper wire R_m, the wire will tear and the process fails. The stress-strain diagram serves to characterise the deformation behaviour of the wire. The deformation properties beyond the yield limit are described by the so-called flow curve. The diagrams in Figure 3.4 are exemplary for copper.

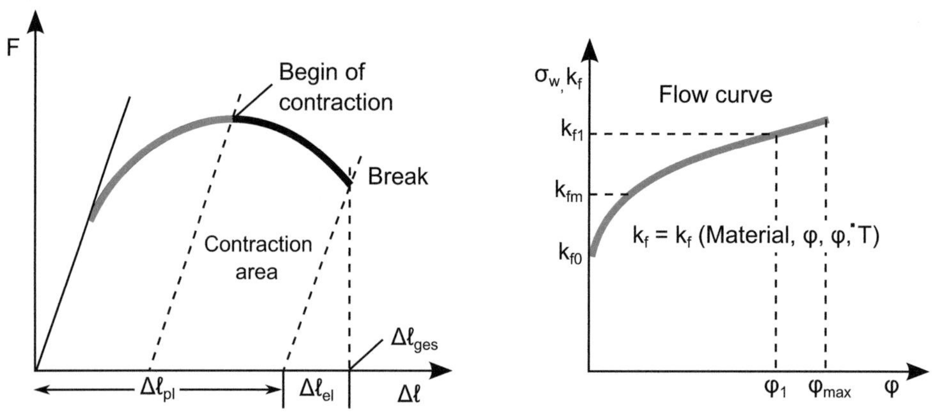

Figure 3.4 Stress-strain diagram of copper wire

Plotted in the stress-strain diagram is the wire tension σ or, respectively, the wire tensile force F over the elongation as a percentage ε. The elongation at break A was introduced in section 2.1.1. The uniform elongation A_g for the elastic region, and the total elongation A_t are also relevant for the deformation during winding. In the flow curve, the yield stress k_f is plotted over the degree of deformation φ. The maximum degree of deformation is labelled φ_{max} and describes material failure.

By winding the wire under a tensile stress, the wire contains a residual tension within the coil after the winding process. Compressive stresses mainly occur in layered structures, where the lower layers are cumulatively constricted due to the residual tension. This is secondary for the actual winding process, but must be considered in the design of the bobbin geometry and the bobbin flange stability. Usually, the compressive stress lies within the elastic region of the wire.

Bending stresses occur when placing the wire on the bobbin radius or the bobbin edge, and when feeding the wire through wire guide in the machine. Due to the necessity for wire tension in the winding process, the bending stress is always superimposed by the tensile stress. This leads to an increased wire tension on the wire's outer radius, which in turn can result in a deformation of the wire profile or to damage of the wire insulation (Figure 3.3). As a result of the wire placement on the bobbin edges, it is plastically deformed. This also applies to circular coils with diameters smaller than 5000 mm.

Aside from a deforming of the wire cross-section geometry, the stresses which occur during the winding process also influence other wire properties.

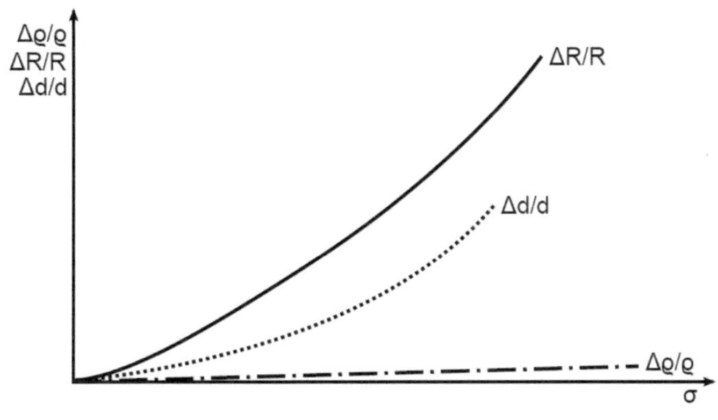

Figure 3.5 Influence of strain hardening and wire tapering on electric resistance [Tzs-90]

Besides the cross section tapering, strain hardening of the material can occur due to the stress variation. This reduces the specific electrical conductivity of the conductor material $\Delta\rho$. However, compared to the resulting wire taper Δd, it is secondary regarding the resistance change as displayed in Figure 3.5. In addition to the increase in the specific electrical resistance, the strain hardening also decreases the elongation at break of the wire, and increases the wire hardness. Accordingly, the wire forming properties, which are important for the winding process, are weakened.

Aside from specifically tapering the wire in the winding process, further deformation effects can be used to influence final wire contours or process forces. An example is counter bending, which uses the Bauschinger effect. This effect describes how bending a wire in one direction beyond the elastic limit results in the involvement of smaller forces when subsequently bending it in the opposite direction. This behaviour is due to repelling forces, which result from dislocations in the material's structure. These dislocations have been built up from the previous bending. An illustration of the strain-stress characteristics is shown in Figure 3.6.

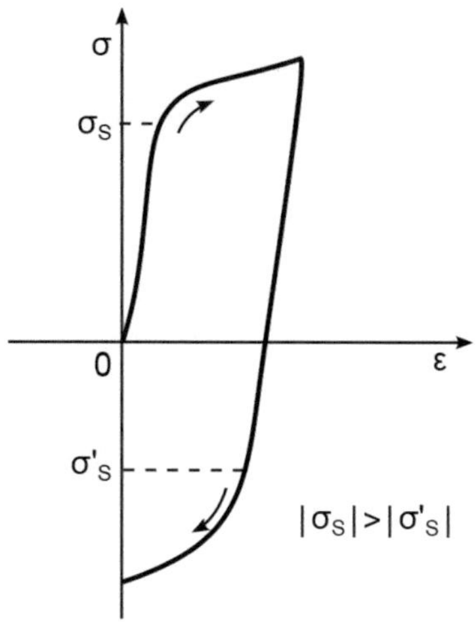

Figure 3.6 Bauschinger effect – counter bending

After the winding process, the prevailing residual elongation and the elongation at break of the wire are crucial for the resulting electrical resistance and the mechanical strength of the winding. These are influenced by the forming properties as well as the process forces, for example, the wire's tensile forces caused by friction.

Based on the wire's minimum bending radii, suitable dimensions have to be chosen for the wire guidance in the machine. The following apply as empirical formulas for pulley radii, wire guiding radii, diameters and minimum bobbin edge radii:

$$R_{roll} = 15 \cdot d_{wire} \tag{3.1}$$

$$R_{nozzle} = 15 \cdot d_{wire} \tag{3.2}$$

$$R_{bobbin} = 1 \cdot d_{wire} \tag{3.3}$$

Additional forming process requirements result from processing profile wires with a non-circular wire profile (Section 2.1.3). When processing flat wire, mainly pulley wire guides with non-round profiles are used, due to their better guiding properties for profile wires. During processing, edge stresses that depend on the chosen bending radius on the long or short side of the profile wire occur. This is why flat wires wound upright usually only result in single-layer coils.

The winding wire is formed by stresses during the winding process. The mechanical processing properties of the wire are mainly defined by its material properties. In the end, the product always has poorer electrical properties due to the wire stresses. The choice of winding process determines the wire stresses. Therefore, the aim of winding technology is to enable a productive manufacturing with the desired product parameters, using the most suitable winding process and intelligent machine components.

3.1.2 Introduction to winding schemes

Depending on the product requirements, for instance a high fill factor or low manufacturing costs, different requirements for the winding scheme can be derived. The most common winding schemes, their properties and applications are introduced in the following sections. As a quantity for the winding complexity the winding index is calculated. It describes the ratio of wire diameter divided by the winding width, times the number of layers.

Random winding

The simplest type of winding, with the lowest requirements for the process, is the random winding. In keeping with its name, the wire is applied to the bobbin in a random fashion. Thus, a high fill factor cannot be achieved. A typical mechanical fill factor for random windings is around 73 % when using round wires. In addition, the diversification of the applied wire length and therefore the coil resistance is relatively high. Random windings are usually created homogeneously by laying the wire with a 1.5 to 3-fold pitch of the wire diameter. The use of this high pitch prevents the slip of an upper turn into a lower layer. This should be avoided because the continuous voltage drop across the wire creates a voltage drop between neighbouring turns. This can lead to an electrical breakdown.

Despite many disadvantages in coil properties, such as high resistance and large installation space, random winding is the most common and most economical winding scheme. The reason is its suitability for mass production. It is characterised by low requirements for the machine and the operator, and the fact that the winding can be manufactured at very high speeds.

The main areas of applications of random windings are contactor coils, relay coils, small transformers and ignition coils. Due to the good production properties, random winding is used in applications with high numbers of turns with relatively thin wire gauges up to approx. 0.5 mm. The following empiric rule applies to estimate the resulting winding height H with a given number of turns n, wire diameter d and winding width B for random windings:

$$H = d^2 \cdot \frac{n}{B} \tag{3.4}$$

A typical structure is illustrated in Figure 3.7.

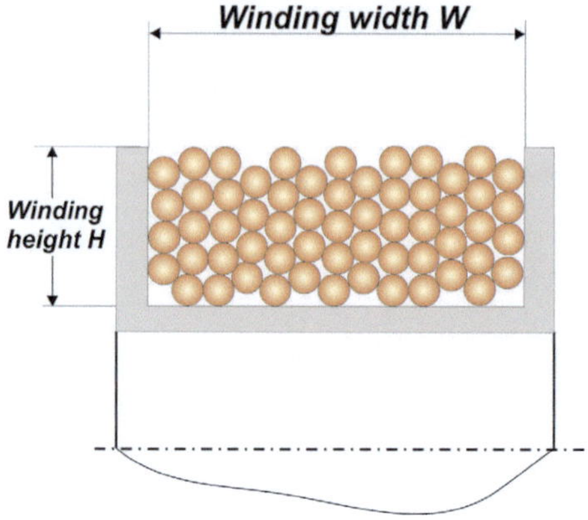

Figure 3.7 Structure of a random winding

Orthocyclic winding

In contrast to random winding, the most challenging winding scheme in terms of manufacturing is orthocyclic winding. The required layered structure can be described as four linear contacts of a turn to the neighbouring turns in the middle winding section. Accordingly, the conductors are mostly parallel to the coil flange. For this type of winding structure, a theoretically optimal mechanical fill factor for round wires of 90.1 % can be created. The aim during the winding process is to lay the wires of the upper layer into the valleys of the lower layer. When placing three circles as the conductor cross section next to each other, it becomes clear that the smallest possible space required is used with an arrangement at 60°. Based on this, the ratio of the available space and the shared space of the circles can be mathematically derived. This is illustrated in Figure 3.8, using the ratio of conductor area to the whole area including a triangle.

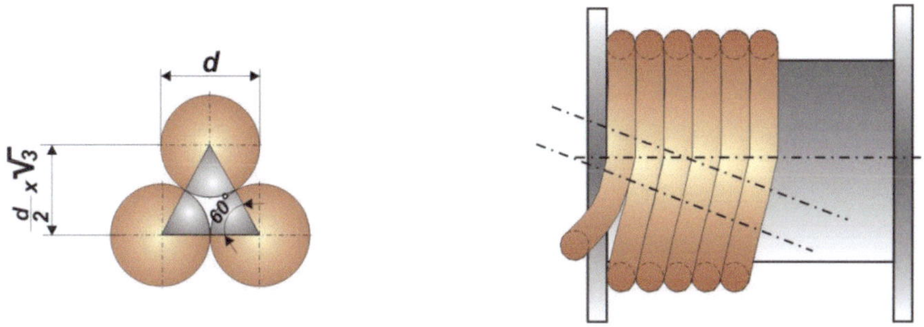

Area of triangle: $$A_D = 0,5 \cdot \frac{d}{2} \cdot \sqrt{3} \cdot d = \frac{d^2}{4} \cdot \sqrt{3} \qquad (3.5)$$

Conductor area within triangle: $$A_W = \frac{d^2 \cdot \pi}{4} \cdot \frac{60}{360} \cdot 3 = \frac{d^2 \cdot \pi}{8} \qquad (3.6)$$

Mechanical fill factor: $$k = \frac{A_W}{A_D} = \frac{\frac{d^2 \cdot \pi}{8}}{\frac{d^2}{4} \cdot \sqrt{3}} = \frac{\pi}{2 \cdot \sqrt{3}} = 0,907 \qquad (3.7)$$

Figure 3.8 Approximation of fill factor for orthocyclic winding

The theoretically achievable fill factor is only valid for the region where the wires assume the geometry as illustrated on the left. For the best use of the available space, the largest part of the winding circumference should be wound in parallel to the coil flange. If the winding then meets the incoming wire, it must perform a turn step the size of the wire diameter. The turn step describes the section of the winding within a layer, at which the wire is lead from one turn to the next. On the right in Figure 3.8, this can be seen for the structure of the first layer of a winding. The incoming wire is displayed on the left coil flange. Due to the regions with a turn and layer step, the actual achievable fill factor decreases. The turn and layer step area takes up between 30° and 60° of the circular coil's circumference. Depending on the wire diameter, the coil diameter, and the winding width in particular, this winding step region can expand beyond a circumference angle of 60°, or respectively on one side of a rectangular coil. If this is not the case, the self-guiding behaviour of the wire is lost and a random winding is produced.

In the end, it is the incoming wire which fundamentally influences the position and quality of the turn step. One must ensure that the wire comes into the winding space at the lowest possible angle. This prevents a wire bulging due to unnecessary diversions, and reduces the necessary space, which is needed for the second turn. In orthocyclically wound coils, the turn step region is always in the same location, in the area of the incoming wire, and continues spirally opposing the winding direction. In consequence a larger coil width, leads to a larger winding step region around the circumference. At the end of the first layer, where the wire is moving up into the second layer, the resulting offset of the turn step leads to a different radial position of the incoming wire for the second layer compared to the first layer. This behaviour can be observed in every wound layer so that a spiral region of crossovers develops for this coil. The typical course of this crossover zone over the coil's structure is shown in Figure 3.9.

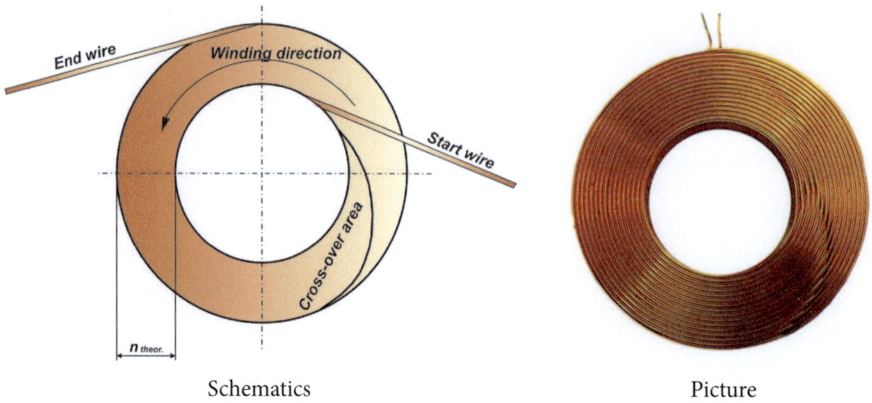

Figure 3.9 Cross-over characteristics

The wire crossings in the layer step region lead to bigger windings heights in this area. This is why orthocyclically wound coils with a circular winding ground are never circular in the last layer. The radially constantly moving layer step causes a hunch-shaped form at this point. Consequently, the winding height at this point is not equal to the number of layers times the wire diameter. Empirical values show that this crossover region adds 5 to 10 % to the actual winding height, depending on winding width, number of layers, and the wire diameter.

Due to the requirement that the windings should be parallel to the coil flanges for as long as possible, it is generally necessary to adjust the winding width exactly to the number of turns in each layer. For non-circular coil shapes, the hunch-shaped form from the layer step should be positioned to the side of the winding head, meaning the short side of the coil. This is because rectangular coils are often used as segments, in a punched full lamination stack or single-pole in a circular arrangement. As a result, the long sides should be slim in order to prevent contact with neighbouring coils or the lamination stack. The basic winding structure can be specified in one of three geometric variants for both circular and rectangular coils. These are displayed in Figure 3.10, with the number of layers w.

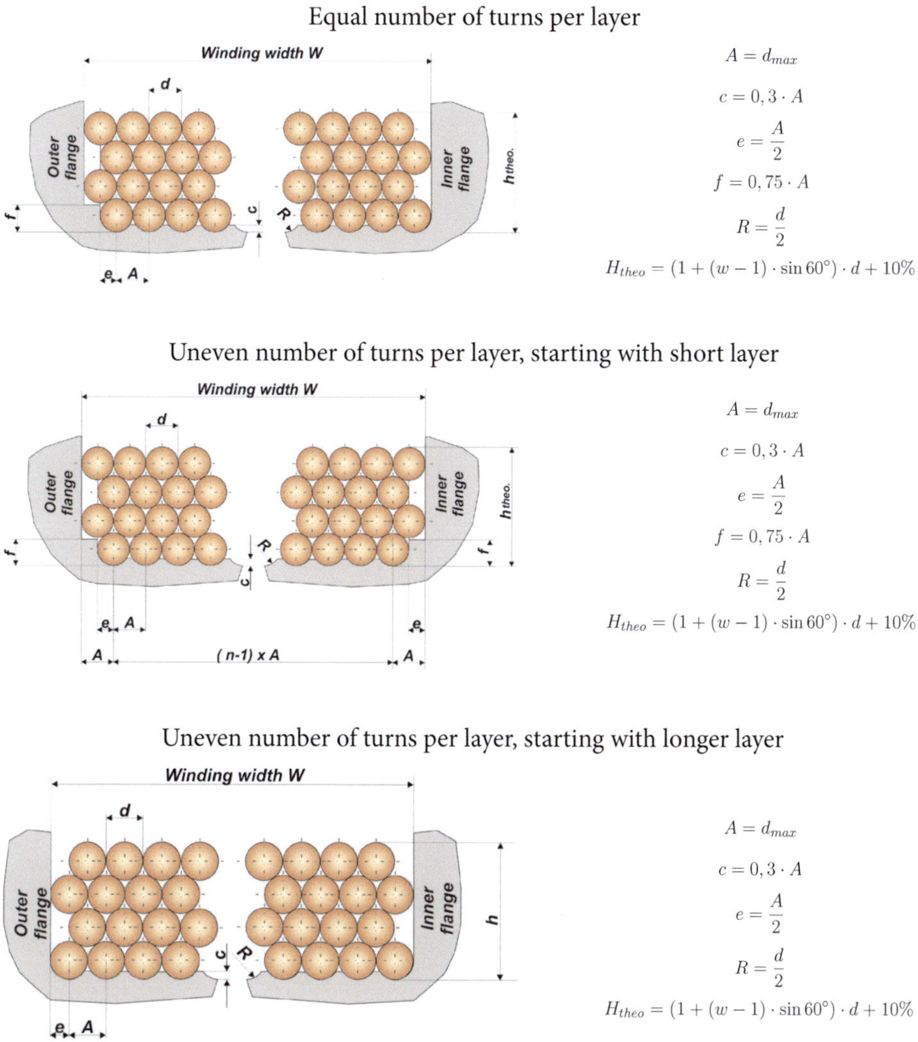

Figure 3.10 Types of layer structure for the orthocyclic winding

The choice of the winding structure used depends on the coil or bobbin design. Among other things, the space conditions available, in terms of winding width as well as winding height must be taken into account. Additionally, it is possible to influence the position of the last turn through a clever choice of winding scheme. The resulting winding height of an orthocyclic winding H is described as an equation in Figure 3.10. Through a parallel orientation of the wire in the upper and lower winding over a minimum of 300° of the circumference, the maximum amount of wires can be applied in the available winding space. Theoretically, a mechanical fill factor of 90,1 % is possible. Although this fill factor is practically not achievable, because of the wire arrangement in the turn and layer step region, the orthocyclic winding is still the best option for making use of the available space for round wires. The manufacturing requirements to create an orthocyclic winding are very high. For example, the sum of all tolerances for the winding window ΔW and the wire diameter Δd need to be very small:

$$\Delta d = 0,4 \cdot \frac{d}{w} \tag{3.8}$$

$$\Delta W = 0,2 \cdot d \tag{3.9}$$

Compared to IEC 60317 this equals around half of the proposed wire tolerances. For orthocyclic windings with profile wires it is possible to put the winding and layer step on the long side of the tooth coil, since the region has the same height as the normal winding region. The necessary guiding structure for the wire in the layering can be achieved by turning the wire over its axis to fit the layer step (Figure 3.11).

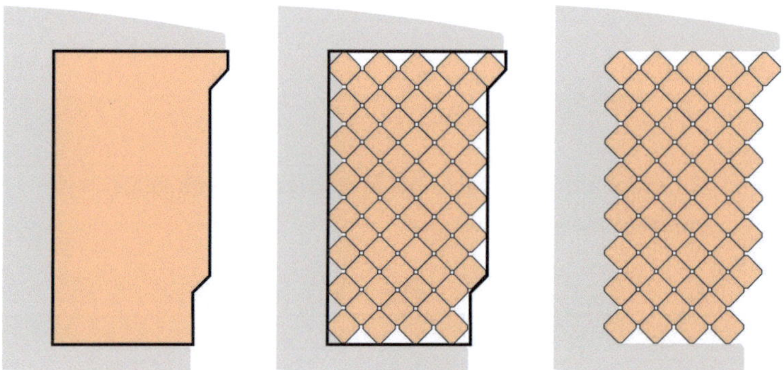

Figure 3.11 Structuring the winding scheme through turning of profile wires

Manufacturing aspects of orthocyclic windings

Even if the requirements for low tolerances for wire and winding space were to be met by adequately high technical efforts, the problem remains that the machine's wire guide must follow the previously described winding structure at high speeds. In practice, this is impossible. If the wire guide would have to perform the turn step for an exemplary 0.3 mm thick wire at 18,000 rpm, it would have only 0.7 s to do so at a speed of 0.42 m/s. At this velocity, the wire position cannot be controlled by the winding machine.

In addition, the problem is increased by the fact that even an ideal wire is practically never straight, but contains a shape deviation due to residual stresses. These dents and bends result from the spooling and transportation of the wire on supply spools. As a result, the wire never places itself straight next to its neighbouring turn in the manufactured winding. Instead, a gap develops according to its dents. The development of theses gaps during winding is also influenced by the surface properties of the enamel layer, like the friction behavior, and the strain behavior, i.e., the copper stiffness. This can lead to a loss of the self-guiding properties of the wire in the winding structure. Typical elongations at break are between 6 and 45 % depending on both wire diameter and supplier. For the manufacturing of orthocyclic windings, however, a high elongation at break is needed.

Due to the limitation in machine dynamics and the varying wire properties in terms of shape and stiffness, it is practically impossible to determine the exact pitch according to the wire diameter for the wire guide. These unpredictable conditions can be compensated by forcing the wire into a predetermined position. The procedure described for structuring the slot ground also leads the wire, so that the wire must not follow the wire guide exactly. This structure is shown in Figure 3.12.

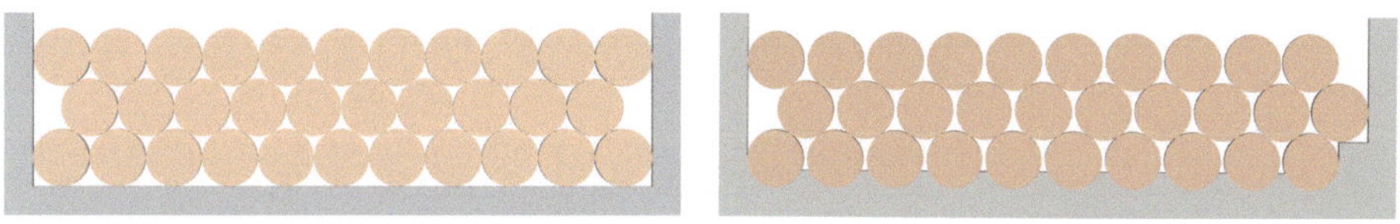

Figure 3.12 Schematics of a structured winding ground on a bobbin

Usually, this grooved contour is also used to ensure the correct number of turns in the first layer regardless of the wire diameter tolerance. The grooved contour also enables the correct wire placement even with a non-digital placement of the turn step, meaning a stepwise and jerky laying motion in the turn step region. Since the wire always loses some cross sectional area caused by the wire tension used and the necessary pulleys (mechanical stresses under tensile load) during winding, the groove geometry is designed for the maximum wire diameter according to IEC 60317. Bends and tolerances of the wire, the bobbin as well as the changing surface properties, can be balanced to a great extent. After the second layer, the wire placement becomes easier since the necessary grooves are provided by the previous layer.

With smaller displacements of the wire guide, the wire guides itself and with the correct number of turns for each layer. For the wire guide motion, one must take into account that the smaller the distance between the wire guide nozzle and the coil, the more precise the movement must be. In return, this means that in some circumstances, the placement motion can be renounced if the distance between nozzle and coil is so big that the force component, as the result of the wire's advance or castor angle, has no effect on the wire's self-guiding properties. Due to the self-guiding behaviour of the wire, is it therefore possible for the wire guide to move continuously, i.e., analogue, rather than stepwise, i.e., digitally, also when using orthocyclic winding. To adjust the correct feed, the angle between the wire outlet at the wire guide and the position on the bobbin is used. This castor angle is a key quantity in winding technology and must be chosen differently for each winding schematic. According to [Dob-11] the following rule exists to adjust the wire's maximum positive castor angle α_{max} depending on the bobbin diameter d_{bobbin} and the wire diameter d_W correctly for circular bobbins:

$$\alpha_{max} = 51,52 \cdot d_{bobbin}^{-0,41} + 11,31 \cdot d_{bobbin}^{-0,33} \cdot \ln d_W \tag{3.10}$$

As a safety factor, 0.4 is suggested for the target castor angle in order to create an orthocyclic winding.

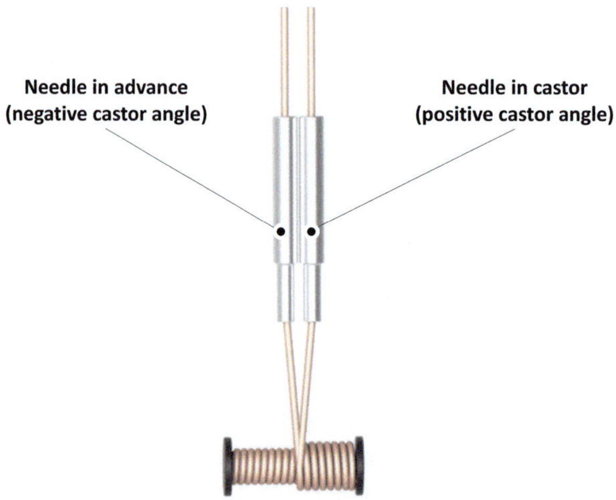

Figure 3.13 Illustration of the castor angle for the wire or wire guide

Helical winding

The helical winding has the geometry of a thread and is characterized by a continuous feed motion of the wire guide in manufacturing. The layer structure features four contact points in the cross section to the neighbouring turns in the middle section of the winding. The wire is not parallel to the flanges, but instead covers the bobbin with a continuous pitch. Even if the upper turn is positioned in the valleys of the lower turns, a perfect fill factor cannot be achieved due to the crossovers. The structure of such a winding is illustrated in Figure 3.14.

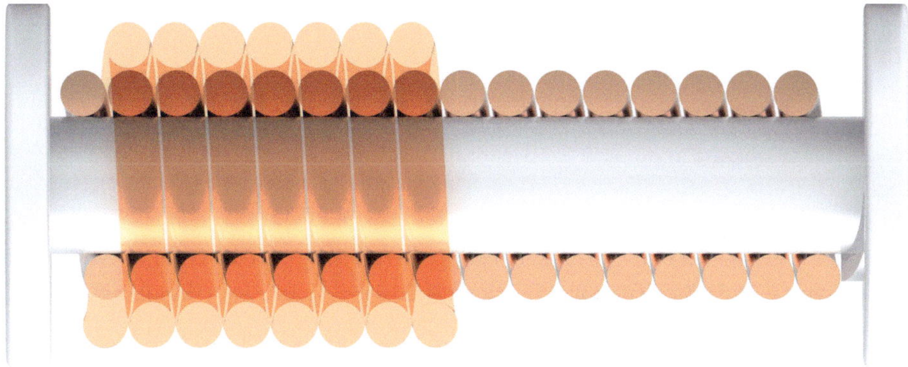

Figure 3.14 Schematics of a helical winding

Typical applications for this winding scheme are single-layer heating coils or rather heating resistances. To ensure manufacturing, often wedges are positioned at the wire inlet (Figure 3.14, bottom right) in order to guarantee the correct position of the first winding. Due to the self-guiding properties of the wire, the following turns will adapt to the given shape by themselves. If the feeding pitch is chosen larger than the wire diameter, gaps will develop in the layer structure, which can be used for further purposes, as insulation distances for non-insulated wires. This winding is called gap winding and will be discussed in this section. By varying the direction of rotation and combining this with a gap winding another type can be realised, which is similar to a double thread.

Besides the wedge for better wire guiding, a grooved geometry can be applied to the bobbin similar to the orthocyclic winding. This measure makes sense especially for high pitches and accordingly larger wire diameters. Due to the continuous pitch of the winding, a continuous feed motion of the wire guide is suitable rather than a digital placement. To estimate the real winding geometry, the ratio of wire diameter and coil diameter can be determined to define the necessary pitch angle. In contrast to round coils, there is a critical angle of 15° for non-circular coils.

Cross-wound winding

The cross-wound winding describes a special layer structure, derived from the thread industry. It is characterised by a spread structure in the shape of a helical winding. An illustration is shown in Figure 3.15. Fill factors are not crucial for these types of application.

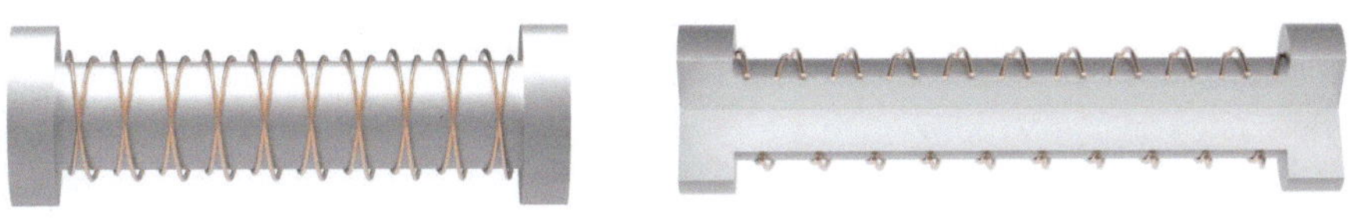

Figure 3.15 Schematics of a cross-wound bobbins

Cross-wound coils made from high-frequency strands are often used in radio technology applications in the medium wave range. A cross-wound coil bobbin often only consists of a winding ground without coil flange due to its low winding height. The winding structure stabilises itself due to the cross-wound wire. By using silver wire, for example, the large surface area of the winding can be used for accumulation and deposition of materials in galvanic processes. The properties relevant to winding technology are the wire diameter, the diameter of the wire bobbin, or the winding. Due to this special structure, it is possible to manufacture the winding in the winding tool and afterwards extract them easily without additional supporting components. Based on the similarities of the winding scheme, optimizing the process capabilities can be achieved by using the same measures as for helical windings.

Pilger step winding

In principle, the pilger step winding is a random winding and therefore has similar fill factors. As opposed to typical random windings though, the pilger step winding creates the winding through individual segments. The structure is shown in Figure 3.16. Typical applications for these windings are pencil coils, or respectively ignition coils for combustions motors. The previously described typical contour results from the spatial limitations of the coil installation. Generally, this winding scheme contains many turns and is therefore suitable for high-voltage applications.

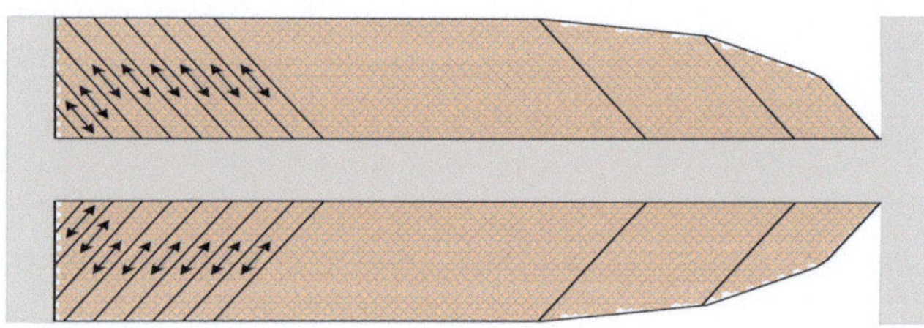

Figure 3.16 Schematics of a pilger step winding

Progressive winding

The winding is manufactured using the so-called *pilger step process* or *pitched winding process*. It features special wire guide kinematics as compared to the processes in Section 3.3. This means that winding sections, or rather segments, with a length of around 2 mm are applied at a maximum angle of 12°. Therefore, the wire guide moves in cycled steps in the direction of the wire placement. The most common windings are made from multiple partial windings of this pilger step process. Pencil coils aim to reduce the potential difference with the step-shaped winding. In order to create these pitched windings, it is first necessary to fill the low-voltage side of the winding space with turns so that a pitched winding surface is created, the so-called start triangle. In the subsequent cylindrical part of the winding, a cylindrical winding section is created with a continuous placement progress. At the end, the wire placement progress is steadily increased so that the winding conically levels off towards the high-voltage side. Similar to reducing the number of turns for chamber coils, this conically reduced part leading towards the high-voltage side leads to improved insulation capabilities due to the decreased electrical potential within the coil. The winding heights of pencil coils using the progressive winding process are typically between 1.5 and 4 mm according to the specifications of the installation space.

Due to the complex coherences between the individual winding parts and many variable parameters, it is often necessary to determine the expected contour with calculation tools prior to programming the winding machine. Therefore, the parameters of the winding are calculated in multiple steps until the desired surface structure is achieved. The quality requirements of pencil coils are similar to those of chamber coils (Section 1.2.4). Despite the pitched winding, individual windings must not slip off. This is why the winding ground is given a certain roughness to ensure that the lower winding has a stable fit on the pencil bobbin. A slip-off of individual turns on the winding surface also creates potential differences, which can lead to premature failure of the system. It is therefore necessary that the wire is processed with constant wire tension during the complete winding process.

Considering that the difference in rotational diameter in a pitched winding within a pilger step can be up to 4 mm, the resulting changes in wire speed are huge. Thus, the wire acceleration at an assumed winding speed of 12,000 rpm can be up to 5 m/s^2. An enormous variation in wire tension is the consequence. These variations in wire tension have a negative influence on the pencil coil quality. For instance, turns can slip off and develop loops, which in turn cause large potential differences within the winding. As for chamber coils, in comparison to Section 1.2.4 (Bobbins), attention must be given to the separation of injection moulding tools. A moulding burr would lead to wire damages and an unstructured winding. Given the nature of the winding process, at each point during a pilger step the wire guide position changes constantly. The wire placed tangentially on the winding surface follows the contour of the winding. This means that the wire exits the wire guide nozzle under a constantly changing angle. Studies have shown that a different exiting angle of the wire leads to a variation in wire tension. However, a variation in tension usually has an adverse effect on the winding quality. To compensate for this effect, the objective to move the wire guide nozzle in three dimensions during the pilger step in order to equalise the distance.

Due to high pencil coil requirements, the dynamics of the wire guide motion must be considered for the winding technology. Inaccurate positioning of the wire guide, or overshooting axle motions, quickly lead to winding errors. The reversing axle motion shows that the motion sequence at the turnaround points consists of a deceleration and acceleration. At a constant winding speed, this leads to an accumulation of turns at the beginning and end. The pilger step winding in particular, is suitable for a monitoring of the wire guide motion using high-speed cameras and adjusting the servomotors depending on the wire guide mass. From the previous points, it can be derived that the requirements of the progressive winding process are high due to the high wire guide dynamics and the varying winding structure.

Toroidal winding

The toroidal winding structure is primarily described by its bobbin. The winding is applied in a rotating process along the circumference. This allows the outside winding to spread out. On the inside, however, the winding has to be more compact. This is conducted with a continuous advance similar to a helical winding. Generally, a structure made of a random winding or a gap winding is possible. A typical design of a toroid is shown in Figure 3.17.

Figure 3.17 Schematics of a toroidal winding

Typical applications for toroidal windings are chokes or transformers. Stating the fill factor is not common due to the applications. The inner core diameter as well as the wire diameter are relevant for winding technology. Due to the winding process, which is explained in more detail in Section 3.3.7, a wire depot must be fed through the opening for an automated process.

Gap winding

This type of winding structure does not per se refer to a fixed winding schematic. Instead, it describes the procedure of creating the winding. For this, the lower winding contains spaces (gaps) to simplify the wire placement in the next layer. This winding structure is used, for instance, in applications with orthocyclic windings and helical windings. A practical example of guiding the current layer on the lower winding is shown in Figure 3.18.

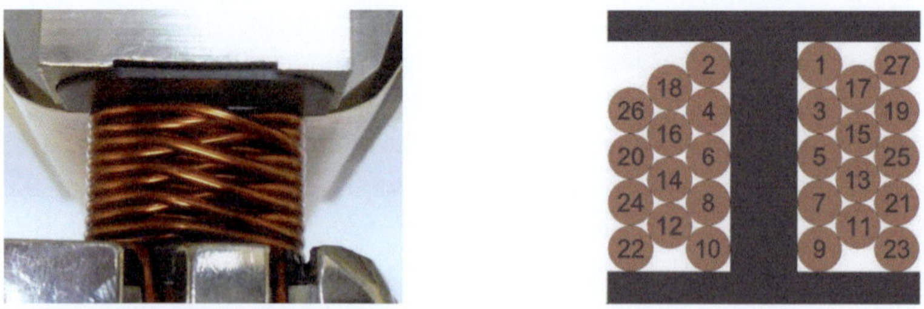

Figure 3.18 Schematics of a gap winding

The achievable fill factor of this winding is slightly lower in comparison to a winding with a complete number of turns in the last layers. Depending on the size of the turn step region this effect is more or less distinct. The winding height is one layer higher than the regular orthocyclic winding in the turn step region. This approach is used when the design of the coil leads to the end wire not being connected to the original flange, or the number of turns in the last even layer is lower than in the previous layers. In the event that the last layer has an uneven order (3rd/5th/7th layer), a spread layer is used instead of a gap winding. Accordingly, this type of winding cannot be assigned to a certain application, because its necessity results from the requested number of turns.

Winding contour

Besides the actual winding scheme, the contour of the winding can also vary, as was shown in the case of the pilger step winding. In general, a convex, concave or conical winding contour is possible, as well as a staged winding. The contour always describes the outer dimensions of the winding. Typical applications for convex coils are single-teeth for motor coils, alternator motors or claw pole motors. The convex shape is especially important for the integration of the thermal sensor onto the single-tooth. For the final assembly of the stator, a conical shape makes sense for single-tooth windings due to the limitations of the winding space. A concave winding is usually not used, instead it indicates a winding error. Exceptions are special shapes, for example, for sensor coils. Staged windings are used in valve coils or tie rods, and usually result from spatial conditions.

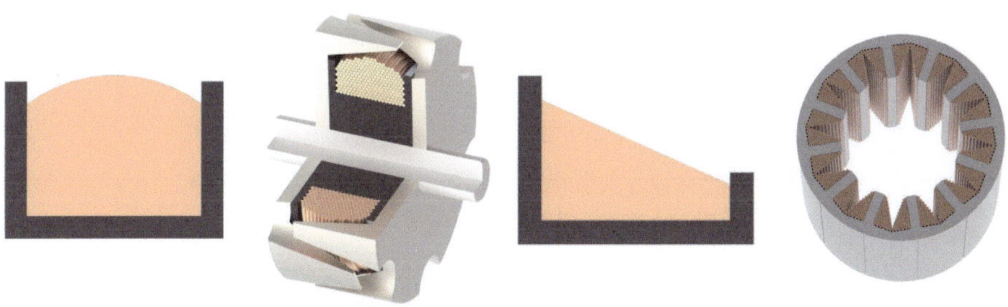

Convex winding contour | Application in alternators | Conical winding contour | Application in single teeth stator

Figure 3.19 Schematics of different winding contours and their applications

Comparison of winding schemes

A conclusive comparison of the different winding schematics is summarised in Table 3.1. The assessment ranges from ++ to --, where ++ stands for the best and -- stands for the unfavorable properties.

Tab. 3.1 Systematic comparison of winding schemes

Parameter	Random winding	Orthocyclic winding	Helical winding	Cross-winding	Pilger step winding	Toroidal winding
Fill factor	-	++	+	--	-	-
Influence of wire gauge tolerances	++	--	-	++	+	+
Suitable for large wire gauges	-	+	+	0	-	-
Suitable for wide winding widths	+	+	+	+	+	+
Suitable for high number of turns	0	0	-	0	-	-
Thermal conductivity	--	++	0	--	--	-
Mech. stability	-	+	+	++	-	+
Suitable for secondary insulation	0	--	+	++	0	+
Requirements for winding process	++	--	-	0	--	+
Requirements for bobbin	+	--	-	++	-	+
Speed	++	0	+	0	+	-
Occurrence	++	+	--	--	-	0

The design and choice of the winding scheme is highly dependent on the product and can therefore not be stated in general. A high fill factor generally has a positive influence on the electrical and mechanical properties of the coil. In Figure 3.20, this is illustrated for the electrical resistance and the inductivity according to a given winding width BK, a coil diameter RK and a number of turns N.

For the choice of wire diameter, the tolerance for the complete diameter has to be considered, because a changed number of turns in each layer results in a disturbed layer structure. In Figure 3.21, a distinction is made between case 1, which describes the orderly layer structure, and case 2, which describes a disturbed layer structure due to displacement of a turn into the next higher layer. As a result, the mechanical fill factor decreases due to the additionally required winding space due to the displaced turn. This also leads to a worse ohmic resistance due to the greater wire length.

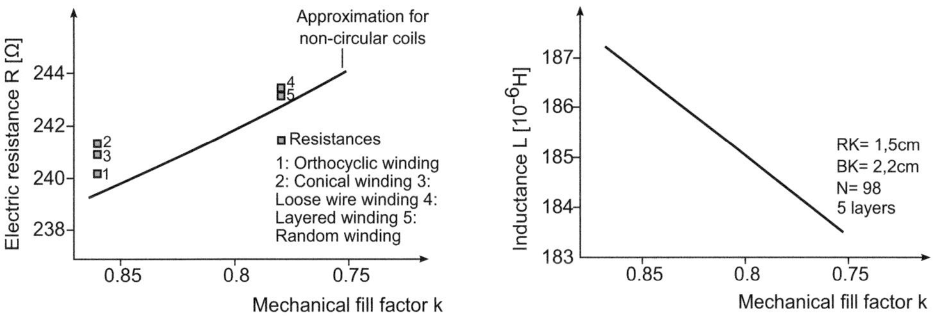

Figure 3.20 Influence of fill factor on electric coil properties [Dob-11]

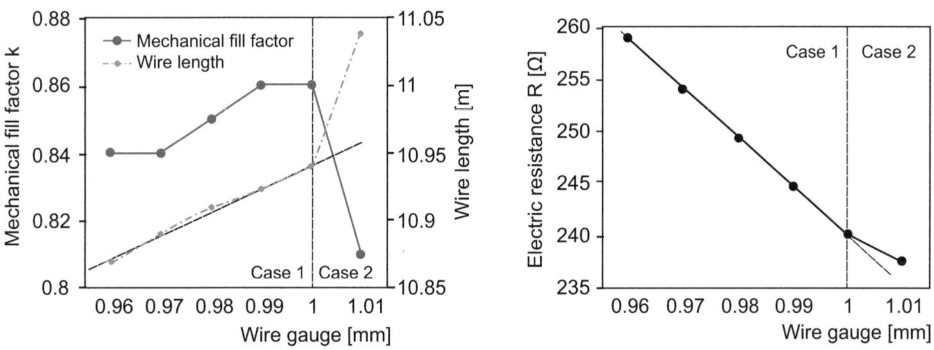

Figure 3.21 Influence of wire tolerance on coil characteristics [Dob-11]

The winding scheme is primarily defined by the functions and requirements of the product. For example, high requirements for power density usually lead to an orderly layer structure. These products are therefore typically wound with an orthocyclic winding scheme, or possibly profile wire. A very short process time is possible for the random scheme due to its low process requirements, and therefore enables higher winding speeds. The requirements of the winding scheme therefore reflect in the process time and the quality of the winding carrier.

3.2 Central functions of winding technology

Figure 3.22 Wire tension control systems in winding technology

As a key quantity in winding technology, the tensile force of the wire during the process has a crucial influence on productivity and product quality. Accordingly, causes for variations of the wire tension and possible counter-measures using wire tension control are introduced in this section.

This section is primarily directed at readers from manufacturing and process development. Wire tension can also be a relevant quantity for the design of possible cycle times.

3.2.1 Wire tension control

As introduced in the previous section, a defined wire tensile force is necessary for a suitable winding process. The mechanical wire strain is used to describe the wire tensile force in terms of the wire cross section. It is defined by a quotient of wire tensile force and wire cross sectional area. Usually, the wire strain is referred to as wire tension. In general, the relative motion of wire and bobbin in combination with the friction on the guiding components in the machine results in a tensile force, which acts on the wire. However, this is much too low for a functioning winding process. Reference values for wire tensions in the winding process are defined in IEC 60317, as elaborated in Section 2.1.2.

Secure wire handling depends on the wire tension. An insufficient wire tension leads to an undefined wire length and position between handling elements. A defined wire placement is impossible in these conditions. If the wire tension is controlled by the wire brake, it enables a flush layer structure on the bobbin, which has a positive effect on mechanical, electrical and thermal properties. A constant wire cross section can be ensured based on a constant and defined wire stress in the winding process (Section 3.1.1). This in turn guarantees a defined coil resistance. The relation between wire tension and the resulting coil resistance is proportional, as is shown for small wire diameters in Figure 3.23.

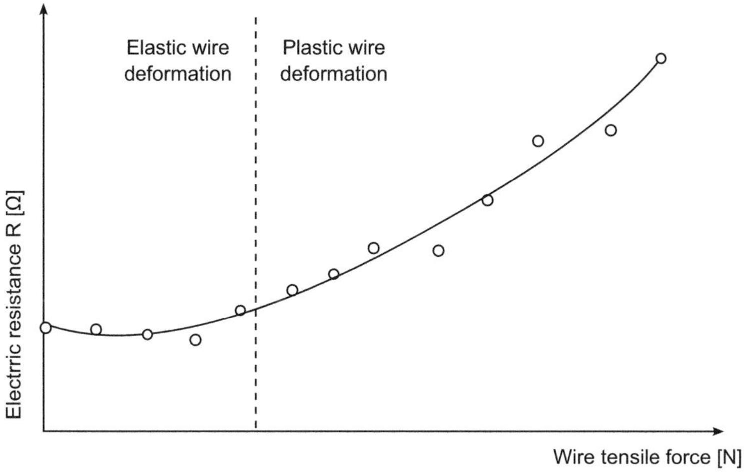

Figure 3.23 Influence of wire tension on electric resistance [Wo-98]

Additionally, a time offset exists between tension measurement and control action. This is why the machine often operates only in an actuator mode without a closed-loop control. The actuator mode guarantees a defined basic wire tension, which is necessary for the winding process, and does not try to control the variation. Usually this actuator mode is complemented by passive compensation elements, to balance the process disturbances.

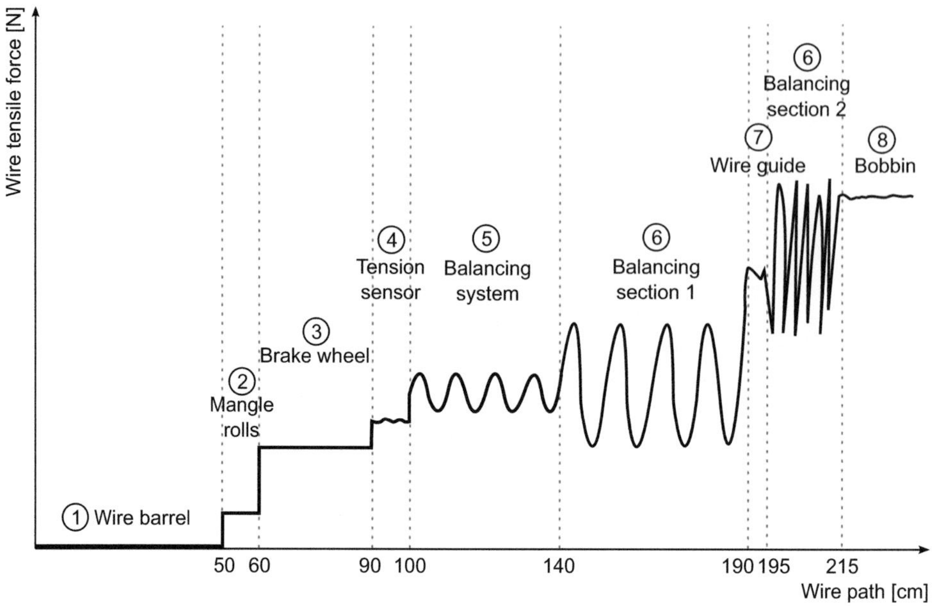

Figure 3.24 Tension characteristics over the wire path

Excessive wire tension manifests itself in multiple ways. A more significant cross sectional taper of the wire, which results in a higher coil resistance, is a clear sign. Other indicators for excessive wire tension are damaged wire insulation, a bobbin deformation or deviations from the desired coil geometry, and in extreme cases wire tearing.

A low wire tension usually leads to loose wire ends in the winding, or errors in the layer structure due to uncontrolled residual wire stresses. In the same manner, this applies for a possible gap between the winding and the bobbin. Higher electrical resistance due to dents in the winding, combined with a deviation of the outer winding dimensions, as well as poor thermal conduction from the winding to the bobbin are further signs of a lack of wire tension. Wire tension control is difficult because of the number of process parameters affecting it. The major influence is the winding speed and the associated wire speed. For non-circular coils, there is a geometry dependent periodic disturbance resulting from the bobbin shape. This scales with the aspect ratio of the bobbin (Figure 3.25).

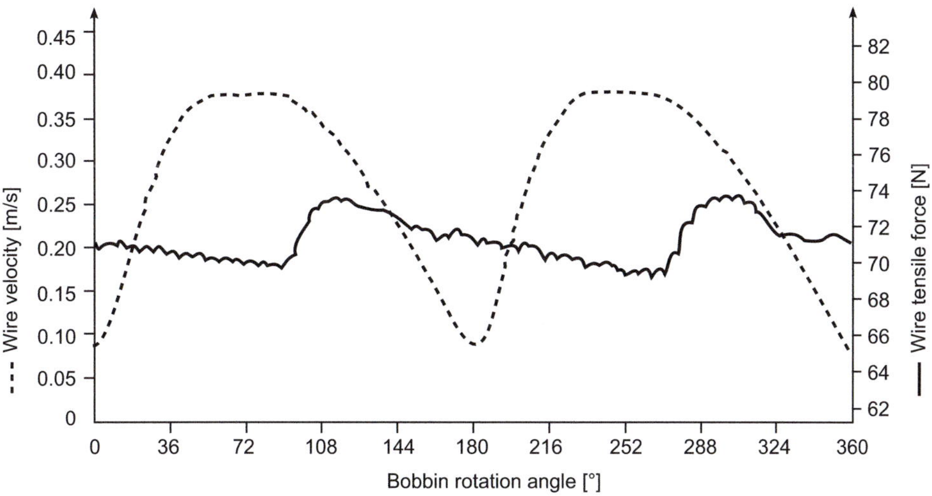

Figure 3.25 Periodic wire tension disturbance through noncircular bobbins (aspect ratio of 1:5): resulting wire speed and wire tension at 60 rpm

The wire also has an influence on the resulting wire tension due to its diameter and hardness, which define the wire stiffness. The friction within the process adds directly to the wire tension but can be influenced by applying of a lubricant. The machine design itself influences the wire tension by the choice of wire guide parameters, like nozzle diameters and nozzle exit radii, as well as the number and the distance between guiding pulleys in the machine.

Generally, there are two force application methods to influence the wire tension with machine components. The first is the most commonly used option of rope friction, respectively general friction force. It can be applied with a brake wheel or brake pulleys, similar to a set of mangle rolls. The second option consists of clamping the wire between friction partners. However, this can possibly lead to wire insulation damage due to excessive pressing forces, which is why it is not often used.

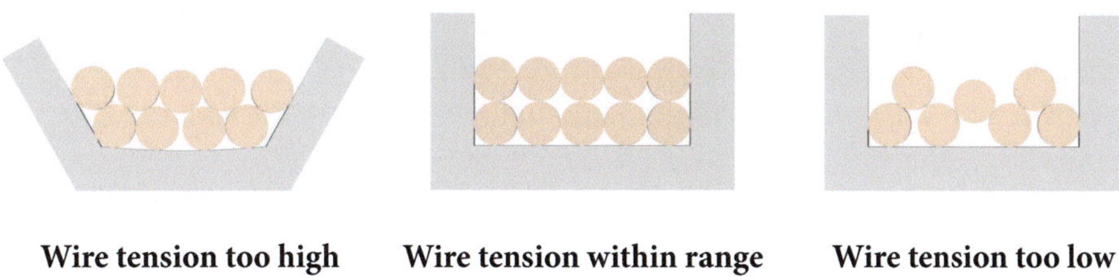

Figure 3.26 Consequences of errors in the wire tension control

Typical actuators for wire braking systems are classified in Table 3.2.

Tab. 3.2 Classification of possible actuator systems for wire brakes

Mechanical systems	Electromechanical systems	Pneumatic/ Hydraulic systems
– Spring system – Slide system (Ceramic disks)	– Magnetic particle brake – Hysteresis brake / Eddy current brake – Servomotors – Linear motors	– Pneumatic cylinder – Hydraulic cylinder

Mechanical systems can be used for a variety of applications but depend on the actual design. Slide block systems, for instance, are only used for very thin wire gauges (up to 0.1 mm). Spring systems, however, can be used for almost all wire diameters but are limited in their tension balancing abilities. On the other hand, the excellent control properties of electromechanical systems have led to a broad distribution in tension control systems.

Wire tension is typically measured with strain gauges, capacitive sensors or optical sensors, in combination with spring scales or rotary encoders. A typical setup of a wire tension control system comprises a set of mangle rolls, which is supposed to straighten out deformations of the wire from the supply spool and already provide an initial tension for the following components. This initial tension is necessary in order to keep the wire on the following brake wheel and to generate friction. For thin wires, a felt can substitute the mangle rolls to create the initial wire tension. A measuring usually roll follows the brake wheel to determine the local wire tensile force. The point of intersection to the subsequent winding process is a balancing system to adjust free wire lengths, also called loose wires. Suitable systems will be introduced in the following section. The typical setup can be comprehended based on Figure 3.27. Two types of brakes are shown, the electronic brake (EB) and the electronic balancing brake (EBB).

The wire tension is a key process parameter in manufacturing coils with a defined winding and the respective positive electrical properties. Since higher winding speeds directly increase the wire tension for non-circular parts, the wire brake properties limit the achievable winding speed for these winding tasks.

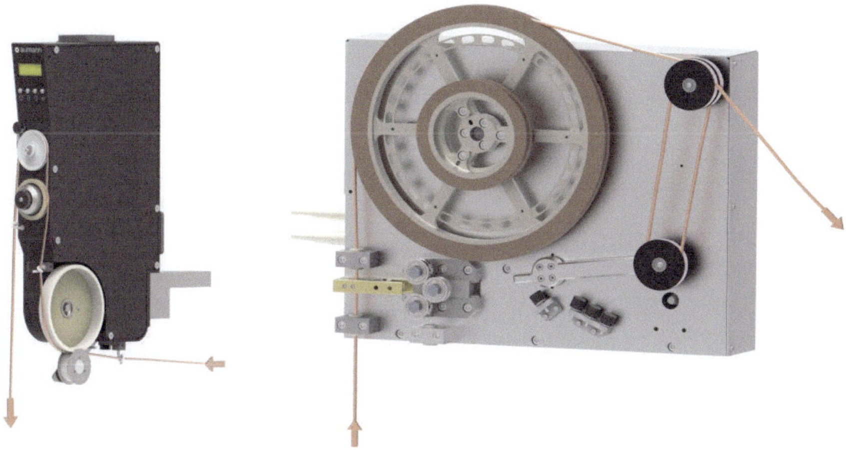

Electronic Brake (EB) **Electronic Balancing Brake (EBB)**

Figure 3.27 Two electronic tension control system designs (Aumann GmbH)

3.2.2 Balancing free wire lengths

Aside from an initial tension application for the winding process, loose wires can occur due to upstream or downstream processes, such as declining wire guide motion, which have to be retrieved by the wire storage. If this function would be neglected, the wire brake wheel would lose its friction and could not control the wire tension. Consequently, a defined wire placement on the bobbin would no longer be possible.

However, loose wires can also occur during the actual winding process. Non-circular bobbins, for example, can lead to significantly longer free wire lengths between wire guidance and bobbin after wire placement on the shorter side during the winding process. As described in Section 1.1.3, start and end wires occur at terminating the wire on the bobbin and must be retrieved for the following handling processes. The wire guidance on the park pin or on the pivoting wire guide are typical upstream or downstream processes which cause loose wire ends.

The necessity of balancing the loose wires can be summarised as follows: A defined position of the wire's end is necessary for the neighbouring processes and also enables a reduction in wire waste due to start and end wire. The wire storage required is referred to as the primary wire storage due to the larger free wire lengths. The secondary wire storage is required for tension balancing during the winding process and provides a constant wire tensile force (Figure 3.28).

A compensation of the free wire lengths can be realised using a cable winch, a dancing lever, a media flow (airflow) or a servo motor wire brake. The latest is the only option, which can also actively feedback wires and therefore has the best control properties. Often a suppression arm, a dancer system or a hoist are used for the primary wire storage. In comparison, higher dynamics and therefore smaller masses have to be ensured for the secondary wire storage. Accordingly, servomotors are used for the active wire retrieval, or pneumatic systems for the compensating.

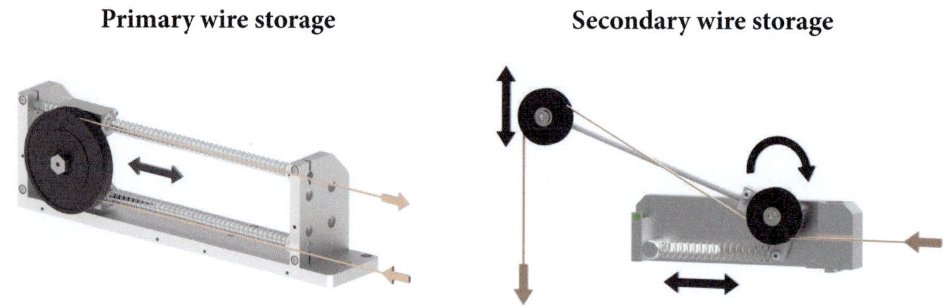

Figure 3.28 Classification of wire balancing systems

However, the use of a direct airflow is only possible for wire diameters of up to 0.1 mm. For larger wire diameters, pneumatic cylinders must be used. An alternative to the secondary wire storages are compensators. These can consist of elastic elements, like lamellas or moulded springs, or rigid objects such as cam disks. These are shown for comparison in Figure 3.29.

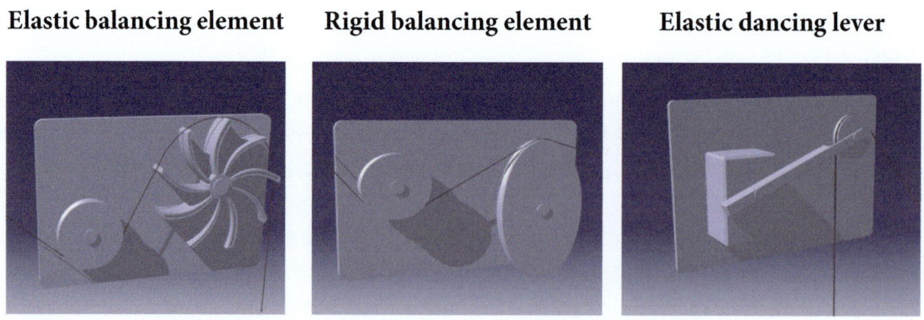

Figure 3.29 Design of different compensation systems according [SLB-15]

The advantage of rigid objects is the very effective compensation achieved by using a simple technology and geometry derived from the bobbin. The disadvantage of these systems is the product specific design and fixed shape, which does not increase with the coil structure in multiple layers. Elastic balancing objects, similar to springs, are usually too inert or too stiff to achieve a similar quality of wire tension balancing.

Different possible combinations of balancing and braking systems are illustrated in Figure 3.27. A type EB brake is displayed on the left. It contains a brake wheel, a measuring point and a balancing system. The balancing system consists of a pneumatic cylinder with a constant operating pressure and spring elements. The control properties of this system are very robust. The initial wire tension can be adjusted very precisely by the brake wheel control. However, the brake wheel can only brake the wire due to the use of a hysteresis brake. On the right (Figure 3.27), a different approach is shown. The type EBB also consists of a brake wheel and a pneumatic cylinder. Contrary to the EB, the brake wheel provides a constant torque, while the pneumatic cylinder can be controlled via a proportional valve. Due to this controllable balancing element, the process variation can be balanced much more precisely. However, the desired initial wire tension has to be adjusted manually through the brake wheel. The control dynamics are better for two reasons. At first, the cylinder can brake using the lever, or release a free wire length. Secondly, the lever mass is significantly lower compared to the brake wheel. Each wire break still has its own justification depending on the chosen process window and the part's properties. Systems with lower dynamics usually have higher forces and can therefore be used for larger wire diameters.

When winding non-circular coils, the balancing system is just as important as the control of the wire tension or the brake wheel, since the balancing system compensates wire tension variations in the winding process. Therefore, the balancing system is often integrated into the wire tension control system. When balancing the free wire lengths resulting from primary wire guiding motion, long wire lengths have to be compensated, since the placement motion is not limited to the winding space. However, there are no dynamic requirements.

3.3 Winding processes

Figure 3.30 Needle winding for concentrated windings

This section covers the key properties and machine elements of different winding processes. After an introduction to kinematics, mechanical engineering including control technology is presented. A methodology for choosing the most suitable winding process for a given product and possible approaches for troubleshooting will be introduced at the end of this section.

This section is relevant to readers from manufacturing and manufacturing planning.

3.3.1 Introduction to winding processes

The purpose of this section is to introduce the basic kinematics of winding technology and the associated implementation in machines. In general, one must distinguish between linear axes and rotational axes for the implementation of movement in a machine. The direction of movement determines the process name. Both types of axis are based on a numeric control technology, also called NC-axes. The winding processes can be classified into a rotation based winding process, a parallel kinematic winding process, a multi-stage winding process and a manual process. Examples for rotation based processes are flyer winding (with a rotational axis and an optional linear NC-axis), linear winding (with a rotational axis and at least one linear axis), and toroidal winding (with two rotational axes). In flyer winding, a rotating wire guide applies the wire to a resting part. In linear winding, the wire is applied to the part via a rotational motion. In toroidal winding, the simultaneous rotation of the wire storage and of the toroidal core create the desired winding. One of the parallel kinematic processes is needle winding, where a cyclic bi-directional rotation of the part is combined with a placement of the wire, using a linear NC-axis. The basic kinematics of the direct winding process can be compared in the following Figure 3.31. Alternatively, the winding can be created in multiple process stages.

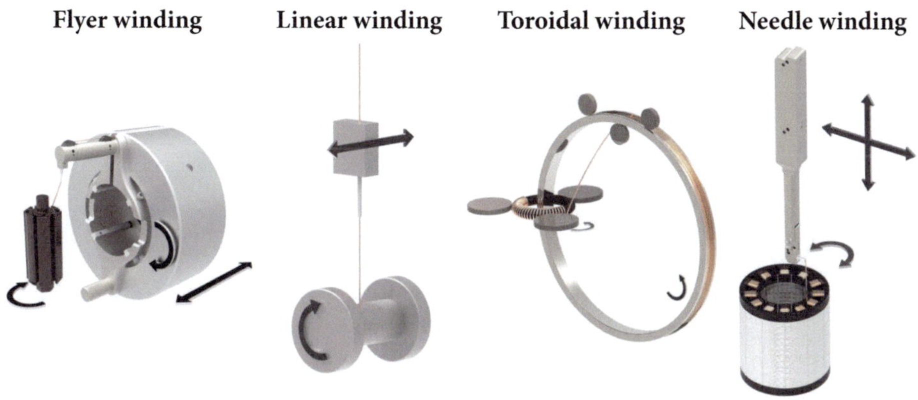

Figure 3.31 Kinematics of winding processes

The multi-stage processes are based on winding manufacturing outside the bobbin and a subsequent assembly. The typical process for this is the insert technology. The coils are manufactured as individual strands for distributed motor windings (Section 1.3.6) on a flyer winding machine or a linear winding machine with multi-stage tools, which are referred to as template winders. Subsequently, they are mounted into the stator slots using an insert machine. This process will be introduced in Section 3.3.8.

The manual assembly process to manufacture motor windings are referred to as fed-in, or trickle winding. The term fed-in describes a process, in which the wires are individually guided into the slot. In fed-in winding, first the winding strand is prepared by winding on a template. After this, the worker feeds the wire into the slot, and the wire position inside the slot is adjusted using auxiliary tools. After placing the wire, the coil strand is continued in the next slot. Lastly, a separate processing of the winding head, the coil part outside the slot, takes place. A systemic comparison of the different manufacturing processes and their properties is summarized in Table 3.3.

Tab. 3.3 Comparison of winding processes

	Linear winding	Needle winding	Flyer winding	Toroidal winding	Insert technique	Trickle winding
Max. speed in turns per min	30,000	2,500	12,000	2,500	-	-
Max. wire gauge	5 mm	2.5 mm	2.5 mm	2.5 mm	2 mm	5 mm
Processing of profile wire	Yes	No	No	Yes	No	Yes
Producible winding scheme	Ortho-cyclic	Ortho-cyclic	Ortho-cyclic	Layered winding	Random winding	Ortho-cyclic
Reachable fill factor	90.1	90.1	90.1	70	60	90.1
Winding types	Conc. W.	Conc. W. Distr. W.	Conc. W.; Outer-grooved: Distr. W.	Conc. W.	Distr. W.	Distr. W.
Wire stress	Small: bobbin	Large: bobbin + needle	Medium: bobbin + flyer arm	Middle: bobbin + needle	Large: assembly	Small
Max. bobbin size	250 mm	500 mm	200 mm	500 mm	1000 mm	5000 mm
Costs	Low	Medium	Medium	Low	High	Low
Degree of automation	High	High	High	Small	Medium	Manual
Typical parts	Mass ↓ Rot. Sym.	Full lamination stacks	Mass ↑ Rot. Asym.	Toroid	Full lamination stacks	Full lamination stacks

In the table the column for winding types includes the abbreviations *Conc. W.* for concentrated winding and *Distr. W.* for distributed winding. Depending on the part's slotting (internal/external), some processes may be used for both types. In wire strains, *Bob* stands for bobbin. This category describes the wire strains caused by the process.

The different processes still share mutual characteristic process parameters. Wire tension, as already discussed, is crucial for all winding processes. Depending on the wire application on the bobbin, different wire speeds from the wire supply may result. These are characterised by winding speed or the number of strokes in case of needle winding, and the related speed ramps for the rotational axis. The angle between wire guide and bobbin is defined as a positive or negative castor angle in the direction of the wire placement of each layer, depending on the position (Figure 3.13). The placement can be done in a digital manner, meaning incremental steps, or in an analogue manner, with a continuous motion. The exit angle from the nozzle, which is perpendicular to the direction of the castor angle, also describes a general process dependency, especially for the resulting wire tension. Lastly, the distances between wire guide and bobbin, or wire guide and wire tension control system, provide information about the required installation space of the machine. The balancing distances have a damping effect on process disturbances and must therefore have a minimum length. Accordingly, the machine's installation space cannot be reduced arbitrarily.

The winding process must be determined and modified for each winding task according to the required product parameters. The boundary conditions from production, including output or cycle times, for the machine must be considered. Almost every winding task can be fulfilled with multiple processes. The suitable process eventually results from the product parameters and the production costs.

3.3.2 Machine elements of winding technology

The purpose of this section is to introduce typical machine elements of winding technology. This is followed by a detailed discussion of winding processes. An example for typical machine components is shown in the Figures 3.32, 3.33 and 3.34.

The typical sub-assemblies of a winding machine are differentiated by their winding specific functionalities. For functionalities, that are implemented the same way in all machines, the same or similar sub-assemblies are used. The wire feeding from the wire spool, for example, takes place over-head from the spool via an unwinder.

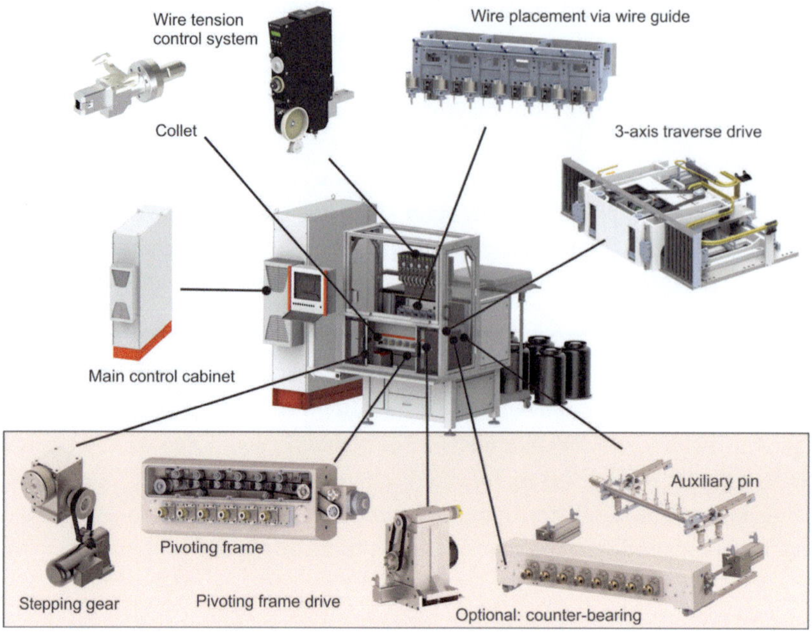

Figure 3.32 Typical components of a linear winding machine

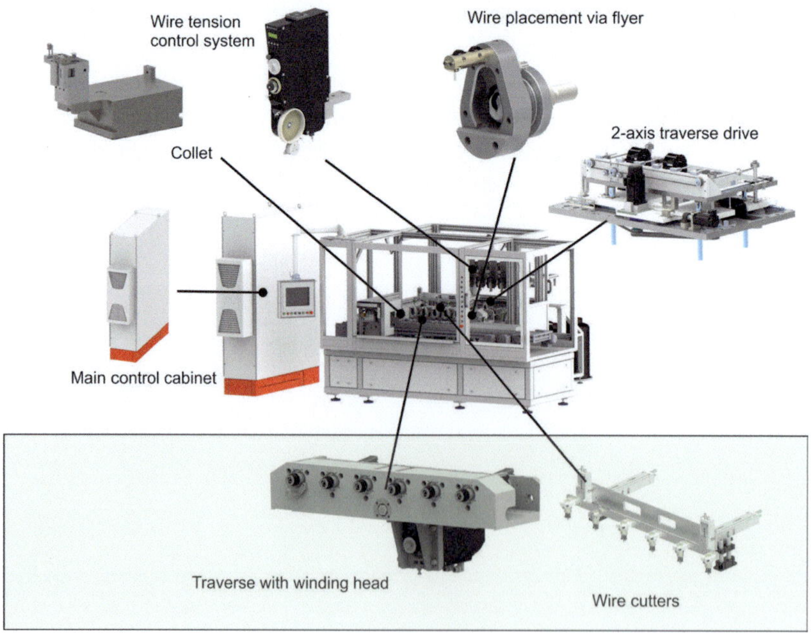

Figure 3.33 Typical components of a flyer winding machine

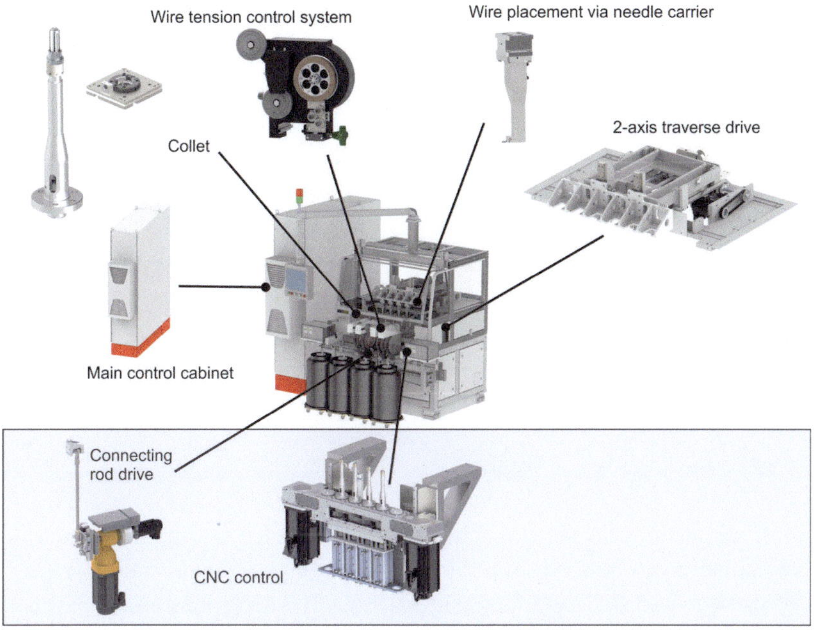

Figure 3.34 Typical components of a needle winding machine

The machine frame, the protection housing and the control is similar for all machines, the wire feeding only differs slightly. Different types and combinations of wire brake and balancing system can be chosen for the wire tension control system, depending on the winding task. For the use of a winding spindle, there is a typical allocation of winding head with housing, spindle bearing and shaft. The linear axes are implemented as servo axes. Lastly, the winding tool and optional wire guiding elements, like park pins or diversion pulleys can be named.

Three typical shapes of wire guides have positioned themselves: the pulley wire guide, the nozzle wire guide and the pocket wire guide. The nozzle wire guide enables an exact positioning of the wire due to its shape. The pulley wire guide is typically used for wire diameters exceeding 0,4mm and enables a more gentle processing of the wire, but, is slightly less accurate in the wire positioning. For slide wire guides, fixed rolls are usually complemented by guiding sheets, which can be matched to the wire diameter, and therefore enable an exact wire guidance in the placement direction. It has similar properties to the pulley wire guide. An overview of the different wire guides is displayed in Figure 3.35.

Pulley wire guide **Nozzle wire guide** **Pocket wire guide**

Figure 3.35 Comparison of wire guides

In most applications the wire must be repeatedly diverted or supported, for which different components can be used. In order to ensure a high wear resistance at the contact points of the wire and the manufacturing machine, the diversions and supports for the wire often use standardized oxide ceramic bushings and guides. Accordingly, their inner diameters must be dimensioned for the wire diameter. The diversion pulleys and their properties also have a special importance, since an undersized pulley diameter has a negative effect on the wire quality due to compression. The pulley radius must be 15 times the size of the wire diameter (Section 3.3). The diversion pulleys should be mounted on bearings.

With regard to the winding process design, including the upstream and downstream process, the cycle time for a component motion is crucial for the design of the overall production line. Typically, the winding process has the highest cycle time in comparison to the other processes, which is why the moving components of winding machines are of special importance. The detailed functions of NC-axes and their design for defined winding are described in pertinent references. Additionally, the winding tasks are highly specific to products as well.

Coil winding technology as a field of special machine construction is characterised by machine technology adjusted for each application. Therefore, aside from components of typical production technology, elements specific to winding are designed and modified for the winding task.

3.3.3 Control technology for winding tasks

NC-axes can be controlled in different ways; either by given movement points, lanes or curves. The type of programming depends on the winding process, although smooth communication of the individual NC-axes is necessary in order to generate complex movement profiles. The hardware concept for the control technology, however, is always similar and is shown in Figure 3.36.

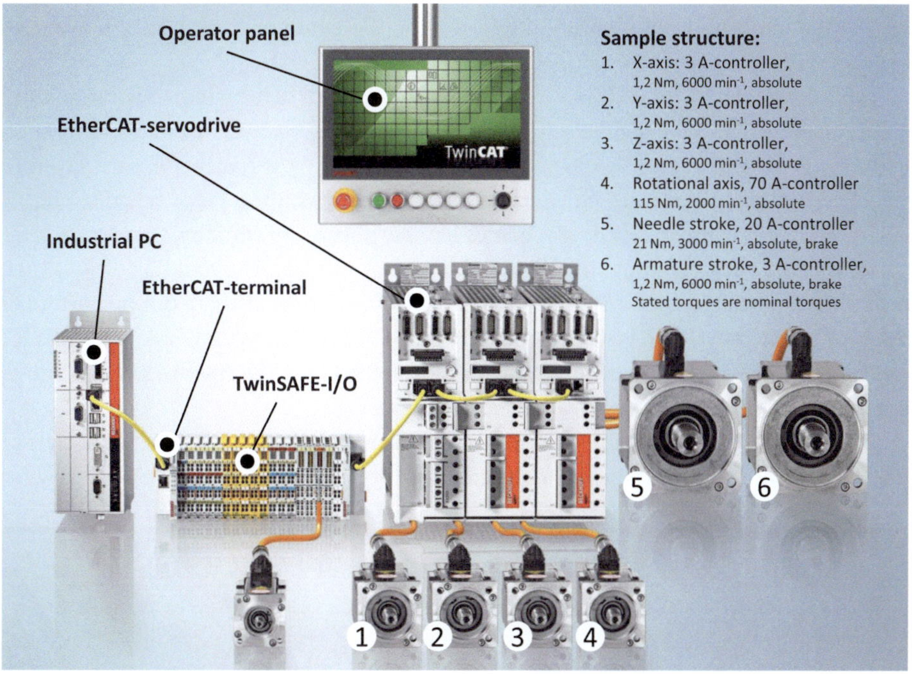

Figure 3.36 Hardware concept for control of winding machines

Communication of the axes

The communication of the individual axes takes place through a control unit which monitors, analyses and processes the axes parameters. Since the wire placement in the winding space should take place as evenly as possible, the rotational axis and the winding head constantly exchange their parameters during the winding process. To keep the positions of the wire guide nozzle and the wound part universal, manufacturing machines usually use servo motor axes. The positioning parameters of the program are continuously controlled by a set-actual comparison. The wire guide motion can be switched between analogue and digital operation. In linear winding technology, the arrangement of the product and the wire guide can be reproduced because of the separation between the wire guidance and rotation of the wound part. Thus, it can be wound on multiple spindles at the same time.

The winding program is divided into different program environments depending on the level of automation and the product variety, but above all, on the operator's qualification. These environments can include different levels with defined degrees of freedom to change machine settings. Each program surface is specific for a winding process, as the winding routine, the set of process parameters to create the desired winding scheme, contains typical process parameters. Usually the winding routine is set for each product, sometimes even based on individual layers. Retooling of the machine with new winding tools for new products is usually done manually, as frequent product changes are rare and the automation would not be economical. During the winding routine, axis movements and their parameters are recorded.

Typical parameters for each layer are: position and width of the winding window, number of turns, pitch or wire diameter with analogue or digital placement, and winding speed including start/stop ramps. A change of parameters can be executed for each layer, for example by moving the winding window, adjusting the winding speed or adjusting the wire tension. For complex winding schemes like the pilger step winding, these parameters are adjusted for each segment (Section 3.1.2). In linear winding, the coupling of placement movement to speed is done with the pitch. In needle winding, a stroke can be performed by a connecting rod, which is why a waiting time for the bobbin rotation must be included. Subsequently, the nozzle can be lowered into the bobbin's slot. The angle of rotation should therefore be as small as possible for a dynamic process.

All control surfaces are similar in their input screens of the basic functionalities, for example switching the machine on or off, start/stop operations or the emergency off switch.

> The objective of control technology is to implement complex movement procedures through coordination of NC-axes. Cycle times, or respectively the achievable productivity, are significantly influenced by the control. For highly dynamic winding processes, programming on the NC-level can be executed without sub-routines.

Linear winding

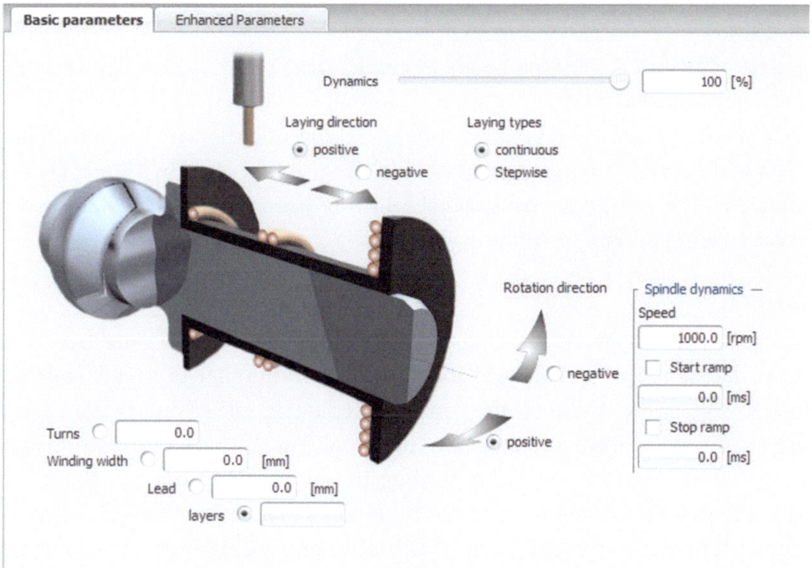

Needle winding

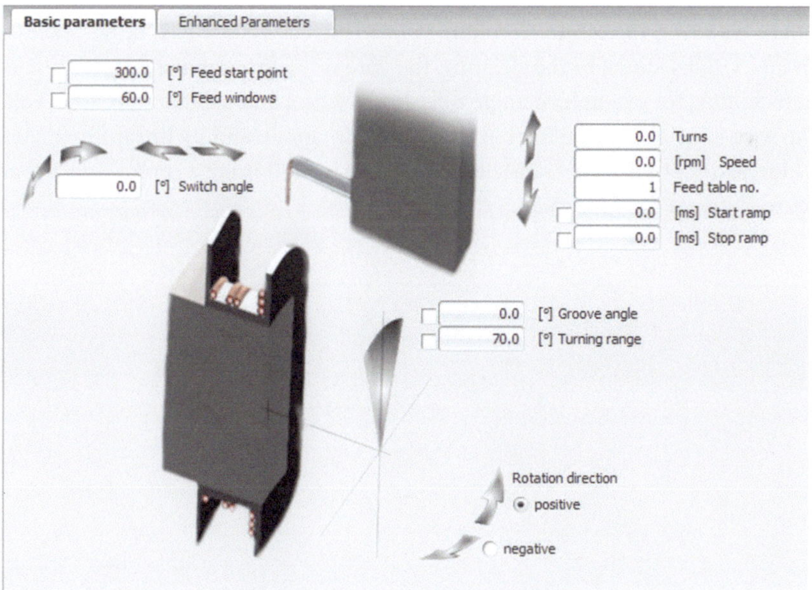

Figure 3.37 Control panel surface for winding routine development

3.3.4 Linear winding technology

Introduction

A large part of electrical winding goods manufacturing requires complex winding tasks that can be performed with linear winding technology. The name originates from the characteristic of wire placement. While the bobbin is rotating, the wire is guided parallel to the bobbin while being moved from one bobbin flange to the other in a linear fashion. The basic principle of this manufacturing technology is to wind an insulated enamelled copper wire onto a rotating bobbin or winding tool.

Wire feeding

Depending on coil design or winding task, the amount of copper wire needed varies. The appropriate amount is made available by a defined number of supply coils near the machine. To supply the machine, the winding medium is either pulled from the supply container, or unwound from a rotatable supply spool. For this type of supply, the wire can amount to 400 kg. For improved wire feeding the supply container can be installed on a jig. This allows wire feeding to the winding object in an orderly fashion, especially for machines with multiple spindles. Immediately after unwinding from the supply the wire can be cleaned.

Due to wire manufacturing with a specific diameter and the subsequent enameling, the copper wire is contaminated with other materials (Chapter 2). There may be remains from storing, for example. In order to avoid dirtying the production line, the wire is cleaned with a felt. However, the lubrication film, which is needed for the winding process, must remain intact. Subsequent to the cleaning, the wire is fed through the wire tension control system including, for example, mangle rolls, brake wheel and balancing system. Following this, the wire is guided to the bobbin using the wire guide and its linear kinematics. The linear kinematics are usually translated into a linear motion by a motor and a gear. This procedure is shown in Figure 3.38.

Figure 3.38 Wire path in a linear winding machine

Motion distribution

In linear winding technology, the wire guide including the wire is positioned to the bobbin and performs the wire placement motion. In contrast to other winding technologies, like flyer or needle winding, it is therefore restricted in its degrees of freedom. Due to the complex kinematics, alternative motion distribution methods, such as integrating the wire placement motion into the rotational axis of the bobbin, are financially unattractive according to current findings. In addition, a movement on the bobbin side influences the motion of wire placement negatively because of the distance between wire and wire guide (Figure 3.39).

Linear winding

Figure 3.39 Illustration of linear winding process. Scan QR code to view video demonstration.

Types of winding machines

There are different types of machines available for various applications. Within linear winding technology, the chosen machine type depends on the bobbin size, its characteristics and the production quantities. The machine variation ranges from conventional table top winding machines (Figure 3.40) to complex and automated multi-spindle machines. The wide variety of machine types available on the market is reasonable, since each machine is designed for a specific production volume. To achieve a highly flexible production for different winding goods, the winding process has already been implemented using articulated robots [Küh-15] or parallel kinematics [Dob-11]. The large number of programmable axes allows for a very flexible design of the application. However, this is only used in prototyping because of the long process times. In comparison, multi-spindle machines provide a much higher utilisation for the same part. The most important advantages of multi-spindle machines are the reduced number of axle drives and the small installation space while maintaining good accessibility, which is achieved by an organised and less redirected way of feeding the wire.

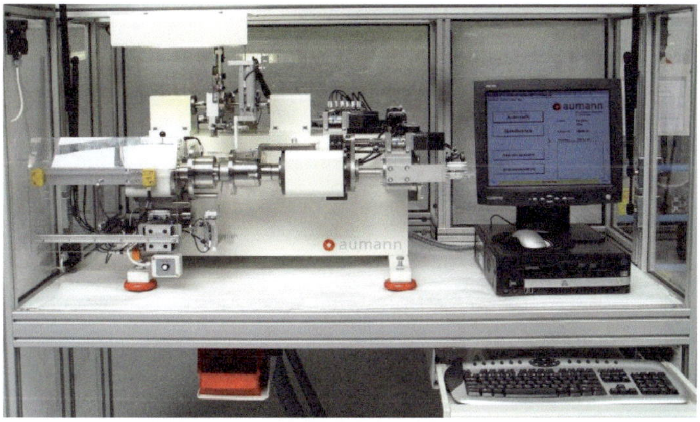

Figure 3.40 Design of a table top winding machine

Manufacturing characteristics

Winding technology for copper wire takes a special place when compared to conventional manufacturing technologies, because it is characterised by two fundamental features. Firstly, the process characteristics strongly depend on the properties of the workpiece. Secondly, the core of winding technology is a forming process from the shape of the wire when stored on the supply coils to the desired shape on the bobbin. With typical manufacturing technologies, such as milling, the influence of the semi-finished product or workpiece on the process time is less severe. Compared to the conventional forming processes, such as forging, bending, stretching, etc. winding technology requires additional special, or rather product specific tools. These distinctive winding technology properties make it impossible to put a machine on the market, which can handle every single winding task. In fact, a number of aspects specific to the product and economical aspects must be examined beforehand, in order to design a suitable manufacturing process. For simple winding tasks and smaller production quantities, there are manually operated table top winding machines. They typically have two motor driven axes, where one performs the rotary motion, drives the bobbin or the winding fixture and the other operates the wire placement motion. The operator's influence on the winding process is substantial for these basic conventional machines. This means that these coil winding processes are prone to error. Because cost distribution between human and technological resources differs on the international market, this technology is often used for mass production in the Asian market. In Europe, table top winding machines have a market for the production of low volumes, or for use in test laboratories. Here, the machines are often combined with special assemblies, such as automated wire cutters or additional motor driven axes. This turns these simple machines into very flexible winding machines with relatively little investment.

The wire guide

The main task for the wire guide platform is to house functional elements, such as the wire guide, that are needed for the winding task. Examples are wire clamps, wire cutters and strippers. The wire guide is the last component touching the wire before placing it on the coil. In the case of a nozzle wire guide, it consists of a tube with the diameter designed for the wire cross section. This is made either from a high-strength tool steel alloy or from tungsten carbide. Therefore its strength characteristics meet the high requirements. The wire guide nozzle has radii on the wire input and output side, and is polished on the inner contour to keep friction as low as possible. High surface properties are needed for a long-lasting and low-friction process. This is valid for all other parts that directly touch the wire or take on an important role in transport.

Bobbin clamping/winding tool

An important part of winding technology are clamping tools, that hold or span by form closure or traction. Typically, the bobbins have geometric properties, that ensure rotational pick-up by form closure. In addition to coil pick-up, the tooling can be used for mechanical stabilisation of the coil core. A collapsing core that reduces the bobbin's constriction (Figure 3.41) is provided for this. Furthermore, a centric position to the winding axis with a minimum unbalance is an important criterion. As the form closure connection is often the most cost-effective, the bobbin design should be influenced to enable a clamping in a winding machine beforehand.

Figure 3.41 Typical design of a winding tool for linear winding processes

Beginning of the winding process

The actual winding process begins with the wire fixation or termination. This is essential for applying wire tension. There are different termination possibilities that are used depending on bobbin, winding task or process routine. General possibilities are attaching a wrapping pin at the coil, or auxiliary or park pin that must be added to the machine or the tool. This starting point of the wire must fulfil important requirements. For example, it has a big influence on the placement of the first turn, as it defines the position of the wire in an inoperative process state, and determines the resulting wire losses/scrap due to its distance to the winding space. For table top winding machines the wire is usually manually fixated on a wire clamp that is attached to the winding tool. This fixating point can be a simple resting mandrel, although a spring-loaded clamp is also technically possible. The right choice often results from the wire diameter and the geometric conditions. However, there are also bobbins with so-called contact posts or coil posts, that are made from the same material as the bobbin. These can have the shape of a fin, on which the wire is fixated. By doing this, the wire is precisely positioned to the coil.

The coil posts can be arranged vertically or aligned to the rotational axis of the coil. For the machine design, this is an important point of preliminary consideration, as the position of the posts must be accurately adjusted with respect to the to the motion sequence of the wire guide. Hence the manufacturing machines need to be adapted to the product, because separate axes must be integrated in order to realise the motion sequences. An advantage of fully automated coil manufacturing is the automatic wire termination at the coil posts. However, the spatial geometries and boundary conditions in the winding space have to be considered in order to avoid collisions due to the complex kinematics. The fixing of coil posts or wire clamping mechanisms is much easier when a flexible traversing motion is possible. A spiral-shaped motion of the wire guide nozzle around the wrapping posts enables a tight and precise fit of the wire. The option of this fully automated process step is enabled by the multi-axes design of the winding machine, which can be freely programmed similar to CNC processing machines. Depending on the machine equipment, this enables a flexible design of the contact pins on the bobbin, but the spatial limitations always have to be considered. Depending on the desired type of electrical contacting process and the interface to the following electrical connection, the coil posts or contact posts are executed as welding hooks, bifurcated contact, insulation displacement contact pockets, pins for arc welding or soldering pins. In several cases, the wire clamping functionality can be implemented with a tool, so that the functionality of park pins is unnecessary when using a wire lock above the wire guide's nozzle in the placement system. In this case, the wire length is exactly adjusted to the functionality of the contacting, so that no wire waste occurs. This is referred to as a wire waste free method. Reasons why the wire waste free method is used include financial benefits, carrying wire remains into the manufacturing machine, and type of contacting. Especially in arc welding or when using welding sleeves, wrapping the contact pin without stranding can sometimes be necessary. Stranding the wire on the coil pin always exists when using a park pin on the machine side, since the connecting wire between coil pin and park pin always crosses the wire guidance in the winding space. When using the strand free wrapping, the start wire of the coil is wrapped after the actual winding process with a special device, which is mostly located on the wire guide system. In this case, there is no connecting wire to a park pin, so that a strand free wrapping of the coil pin is possible.

Park pins

An additional component on the machine is the wire park pin which holds the wire. In continuous manufacturing processes, which must consider product changes, or for bobbins without contact posts, which are directly wound on winding tools, it is necessary to provide separate wire clamps or additional components for terminating on the tools. These pins can either be spring-loaded clamps or similar to posts, which wrap or clamp the wire similar to the termination process on the coil. During mechanical design, it must be kept in mind that the winding space is not obstructed, since the wire park pins, like all other elements, cannot disturb the actual winding process. However, they must be reachable for the wire guide, as illustrated in Figure 3.42.

Figure 3.42 Usage of park pins within the machine

Fully automated winding process including terminating

In fully automated manufacturing, the wire is fixated on a wire park pin, which is the starting point of the winding process. In the next step it must be connected to the bobbin or the winding tool. As previously mentioned, either the bobbin contains contact posts, or the tool features pins or clamps. After terminating on the provided terminating devices (e.g. pins of clamps), the wire must be separated between the two contacting points. Before the actual winding process, the wire park pins must be moved or turned away from the manufacturing area. The resulting wire waste can be disposed at the same time and therefore material is lost. The device must be free of wire for the next termination. Next, the wire is stranded on the contact or coil post and is wrapped towards the winding space. In fully automated manufacturing, the winding process can now begin. The wire contact and the exact positioning of the wire guide nozzle define the placement of the first turn. The following placing can be influenced by the wire guide's linear axis. After the winding process, the wire is fixated in reverse order on the coil contact post or on the wire clamp located on the tool, and later on the wire park pin. The wire is separated again between coil and wire park pin and the cyclic process is back at its starting position. The created coil can be extracted by an automated handling or manually, and a new bobbin or an empty winding tool must be fed to the machine again.

Manual manufacturing

The operator of a table top winding machine faces the issue of having to handle two loose wire ends. After terminating, the end wire often unwinds itself from the coil, while the created winding fixates the start wire. Aside from wire handling, the finished product must be extracted from the machine. Often additional process steps are needed in order to extract the coil from the machine without any quality losses. The end wire is often secured with tape.

Wire clamps or wire locks on the wire guide system

A wire clamp or lock is necessary when directly wrapping or clamping the wire on the bobbin or on the winding tool. Due to the separation of the wire after the winding process and the necessity for a constant wire length for wire termination, it needs to maintain a certain tension. Therefore, it must be fixated between the winding processes. This clamp should be positioned as close to the wire guide nozzle as possible and must ensure a firm grip on the wire. In most cases, the clamping during the winding process is executed pneumatically. During inactive phases, it holds the wire using a spring force as shown in Figure 3.43.

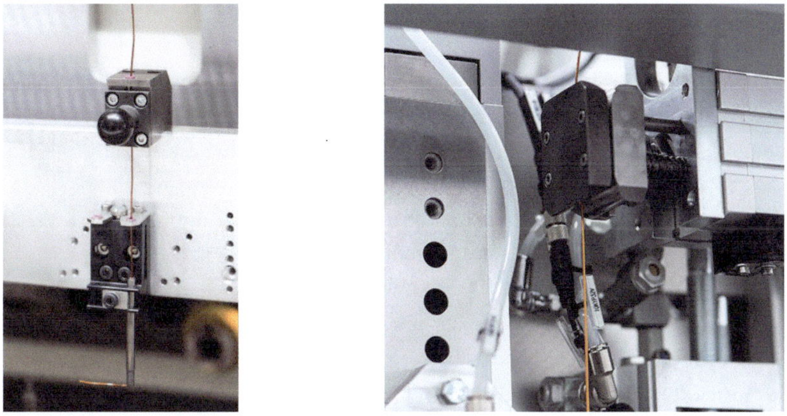

Figure 3.43 Usage of wire clamps

Wire separation processes

The different manufacturing requirements demand the use of diverse wire separation processes. The wire separation processes are distinguished between wire tearing and active wire cutting. The active wire cutters can be pneumatically or electrically operated cutting blades, which are standardised for a number of applications. Enamelled copper wire can usually be torn up to a gauge of around 0.3 mm. The tearing takes place by moving a tear pin, or the wire guide itself, closely past the coil post and separating the wire with the movement (Figure 3.44).

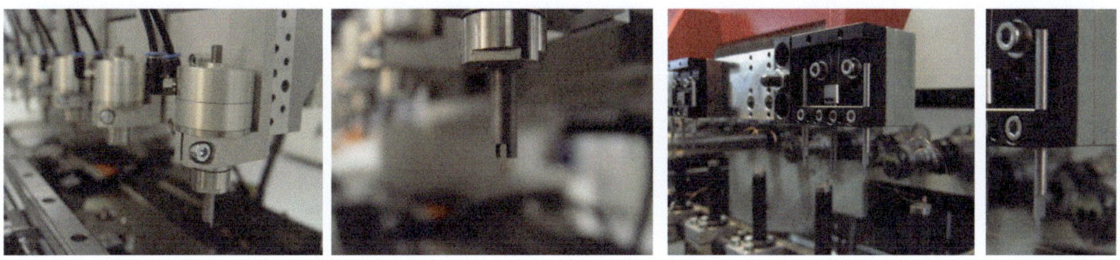

Figure 3.44 Usage of wire tearing units and cutters

The separation point should be close to the coil post, in order to simplify the following contacting process (soldering, welding, etc.). Wire excess lengths should generally be minimised, as they may later lead to disturbances in the product operation. Many special solutions are possible. However, it must be considered that auxiliary tools, for instance crimping tools or wire cutters must be fed to the bobbin, while at the same time ensuring sufficient space to the actual winding process. Parts which perform the additional processes cannot interrupt the motions sequences of the winding process. These additional wire guide system elements are often flexibly arranged in order to keep them out of the operational area of the winding process.

Start and end wire

In many applications it makes sense to have access to wire park pins and wire cutters on both sides of the wire guide. For a continuous wire processing it is important to position start and end wire close to one another, as the coil winding end position is often the same as the start position. Accordingly, preliminary considerations for the position and design of the wire park pins and cutters are appropriate. In some applications these components must be arranged on both sides of the coil, as otherwise a subsequent process may not be possible. In addition, after the winding process, the winding products are processed on the terminated contacting points, or on their start and end wire. These consecutive manufacturing steps have to be considered. For a subsequent welding, wire waste must be avoided. However, additional wire lengths are a basic requirement for wire stripping which describes the removal of the insulation layer from the wire.

Manufacturing of air coils using linear winding technology

For bodiless or air coils the wire is wound directly on a winding tool. With the use of an adapter piece, a winding mandrel or winding tool, different coil geometries can be realised and adapted to the drive spindle. For all fixtures and interchangeable adapters, a rotational pick-up of the part and a centrically mounting position needs to be ensured. Either form closure or traction pick-ups can be used. During the winding process, air coils can be ventilated with hot air. They directly self-bond during winding and subsequently have sufficient mechanical strength for handling. Alternative processes are electrical bonding or oven bonding. The term *bonded coil*, however, gets its name from the original process. Hardening the enamel can also be performed after the winding process. When bonding during the winding process, the force resulting from the airflow can be critical for thin wires during winding. This is why a suitable arrangement of the air nozzles to the winding process must be considered in order to ensure an evenly high winding quality. This and further processes are discussed at the beginning of Section 3.4.3.

Counter bearings

Counter bearings are often needed to stabilise the bobbin and to take up wire placement forces. As the wire is placed with a defined wire tension, the bobbin or the winding mandrel experience lateral pulling forces. Therefore, it is beneficial to provide a counter bearing for consistent quality for the whole traversing length. In case of large outreach of the bobbin on the drive spindle side, a counter bearing creates the required stiffness and therefore stability of the winding process. The setup is illustrated in Figure 3.45. Counter bearings are a necessary sub-assembly for the manufacturing of air coils. As the coils are built according to a pre-defined width, and the winding mandrel is designed with a sidewall to place the wire, a second sidewall, which is located on the counter bearing, must be provided. After coil winding, the counter bearing is retracted and the finished coil can be extracted from the winding tool. The counter bearing can also be part of a winding device, especially in manually operated machines. In many applications, the inner coil diameter is so small that a pick-up of the counter bearing cannot be guaranteed, so that both sides have to be driven synchronously. The counter bearing often engages with the form closure contours of the bobbin when winding. However, the counter bearing is not only used for air coils, but is also used for other bobbins. Similar to a turning machine, the counter bearing acts as tailstock.

Workpiece carrier as counter bearing

Counter bearing in the pivoting frame

Figure 3.45 Design with counter bearing

Pivoting frame

For high capacities of multi-spindle linear winding machines, pivoting frames with tool fixtures on both sides can be used (Figure 3.46). Due to the use of pivoting frames, the extraction of the finished coils and the subsequent equipment with new bobbins can be performed during the winding process for the unengaged coil fixtures. As a characteristic, the repeat accuracy determines the quality of the winding results, which is essentially defined by the frame pivoting gear.

Fully automated linear winding machine with 20 spindles and pivoting frame

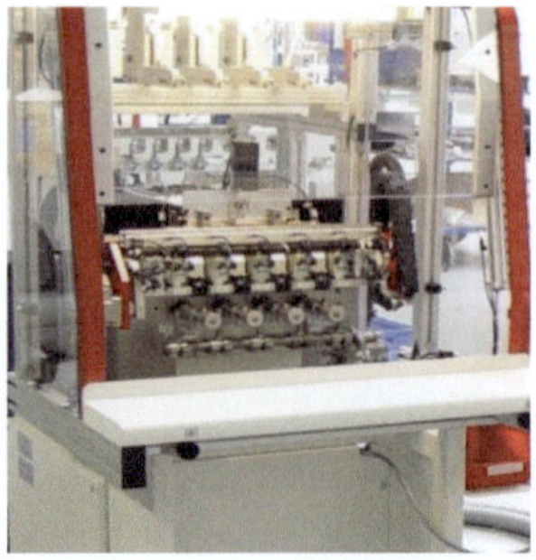

Linear winding machine with 4 spindles and fixed frame

Figure 3.46 Designs with and without pivoting frame

Influences on the manufacturing qualities of coils

A high manufacturing quality of the tools is the prerequisite for the high quality of the winding goods. As all motions during an automated winding process are performed by CNC-axes, it is possible to manufacture both orthocyclic windings and the more common random windings. In addition, special winding geometries can be manufactured due to the flexible machine design with motoric axis. Precise winding fixtures for bobbins and constant repeat accuracy for adapters and interchangeable fixtures are prerequisites for high quality winding goods.

Wire placement for different coils

With linear winding technology, both circular and non-circular coils can be wound. The wire placement on a cylinder-shaped circular coil is very constant due to the evenly increasing gauge, which results from the constant filling of the coil space. It can therefore be placed in an orderly fashion with a sensitive placing gear, a structured program and a constant wire tension. In contrast, the wire tension control plays an important role in the manufacturing of quadratic or rectangular coils. The long distances on the surface sides and the very short distances on the edges cause a speed variation during the coil rotation. These have to be balanced by the wire tension control system with a suitable dancing lever (Section 3.2.2). The design of each system is specifically adapted to the application, as the maximal wire strains are different depending on the wire gauge. The required quality is the key criterion for the design of the process and the wire tension control. Ultimately, the alternating strain is very different, depending on the geometry and structure of the bobbin. If the wire strains are too high, the winding speed is reduced in order to maintain a stable process.

Rotational speed

The main motion for linear winding, which is the rotation, originates from the coil itself. A linear placement motion of the wire guide creates the wire allocation in the winding space. The traverse axis of the wire guide is moved with regard to the rotational axis, and depending on the wire gauge, so that a placing pitch results. High speeds, especially in the processing of thin wires, can achieve up to 30,000 rpm. As a result, there are distinct requirements for carrying components (e.g. winding head) of the drive spindle, since vibrations resulting from the material and the design must be damped. This is why materials like cast iron with lamellar graphite and aluminum cast alloys for weight reduction are used. The high dynamics require a modification of the machine design to the product specific process. Wire speeds of up to 30 m/sec can occur, depending on the winding gauge and the winding process. The winding fixture and the wire guide components are designed as interchangeable parts, so that a linear winding machine can process different bobbins with regard to reasonable re-tooling times. However, several parameters of the whole machine must be considered. Important parameters of multi-spindle machines are the axial spacing between the individual winding spindles, and their number. Key indicators are coil size, winding quality (orthocyclic winding) and wire gauge, which are especially associated with the number of turns.

Special technologies in linear winding technology

A special feature is the simultaneous placement of multiple wires. A distinction is made between parallel wire, bifilar wire and double-wire or multi-wire technology. Parallel wire winding describes the winding of two wires on two separate coils without intermediately parking the wire. Bifilar winding describes the simultaneous winding of two wires on one coil or one winding space without parking the wire, but with separate termination. Winding two or three wires subsequently that involves parking the other wires while one wire is wound, is referred to as double- or multi-wire technology. For these manufacturing processes, the guiding elements and diversion have to be designed accordingly. The winding technique for two or multiple wires needs wire guides for each wire (Figure 3.47). The enamelled copper wires can have the same or different gauges. The most common application is the transformer winding.

Other applications are the processing of coils with twisted or stranded wires for wire terminating at the start and end of a winding. This shape is achieved by a specific twisting technology between the tension control system and the wire guide. The wire guide itself needs to be adjusted as well.

Bifilar technique

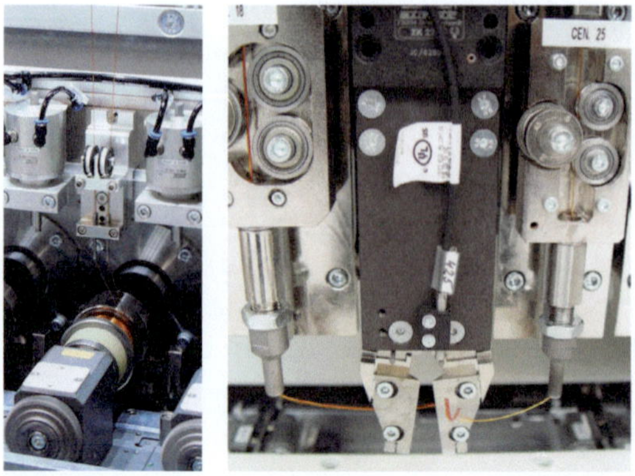

Multi-wire technique

Bifilar winding **Multi-wire technique**

Figure 3.47 Processing of several wires. Scan QR code to view video demonstration.

Bandaging, inter- and cover insulation

The requirements for coil manufacturing often also demand the electric insulation of coils. Bandaging the coil can take place as an intermediate step during the winding process, or at its end. To realise this manufacturing step, in many cases sub-assemblies are integrated into the wire guide carrier. A separate installation of the sub-assembly is possible, but must consider the wire guide's degrees of freedom. Otherwise, the bandaging unit is installed as a separate sub-assembly in the winding machine.

Setting gauge

Wear parts, but also axial positions can quickly and safely be set with the use of setting gauges. Since it subject to wear, the wire guide is the part that must be replaced most frequently in linear winding machines. To avoid adjustments in the program, nozzle stops are provided. If these cannot be integrated due to constructive or financial reasons, setting gauges are used to set the nozzle positions.

Maintenance and cleaning

As fully automated winding machines implement different kinematics, in most cases several guiding units are installed. The stroking motion of the individual axes is usually created by ball screws. As in conventional processing machines, these must be lubricated and maintained. Due to the use of gaiters, the guiding units in series machines are often protected from external contamination. Every part that is in contact with the wire is subject to wear. To avoid contamination of the machine, the wire should be cleaned from excess paraffin or dust particles before it reaches the tension control system. For technical reasons, the supplied bobbins should be clean before being fed into the machine. As winding technology is a chip-less manufacturing technique and there is no abrasion of the wire, it is a very clean technology and can principally be used in a clean room.

The linear winding process is a very rational process due to the rotation of the part and the simple wire handling. The cycle time to manufacture a component results from the quotient of winding process time and the number of spindles used. Fully automated linear winding machines with multiple spindles can often be found in rational applications, where low-mass bobbins are wound, which are rotationally stable, and a high utilisation of multiple winding spindles is required.

3.3.5 Flyer winding technology

Flyer winding technology

The name of this technology is derived from the motion sequence, or rather the wire behavior, during winding. Similar to the first spinning machines, the winding machine adapts a fast rotational motion for the winding of the part. The bobbin is fixedly mounted and the wire carries out the rotation in the shape of a circular motion around it, guided by an extension arm or flyer. This characteristic mainly sets this process apart from the linear winding technology. Typically, the placement motion is done by the flyer winding head which uses a motoric axis.

Flyer winding

Flyer winding of distributed windings

Figure 3.48 Introduction to flyer winding. Scan QR code to view video demonstration.

From the history of flyer winding

This process technology originates from a process with consecutive process steps, also referred to as follow-up winding. For decades, it has been set up with a rotary-table winding process with flyer winding head. The rotary table is a disk which performs a defined rotational motion using a motor. The winding tools are attached to the rotary disk, which is centrically mounted on the rotary table. Wire clamps may be located between the winding tools. They are fixed to the rotary disc. Before starting the first winding process, the wire must be pinned to the wire clamp which is lying in front of it (facing the direction the table is rotating in). The process begins with the flyer head's move to the manufacturing position, which lies at the coil beginning of the manufacturing part. The winding of the bobbin starts with the rotational motion of the flyer. Due to the fact that the winding fixture including the bobbin is rigidly mounted on the rotary table, the winding head must perform the placement stroke for a defined wire pitch. An independent placement motion of the coil is not suitable for this technology due to the constant rotation of the rotary table. Additionally, the power supply to a cycling rotary table is complex, prone to maintenance, and cost-intensive. After the manufacturing process, the end wire of the winding is dragged along by the table rotation and can be fixated by the wire clamp which is actuated by an external cylinder. Consecutively, the rotary disk including the next winding tool is ready for the flyer head. A simple axial motion towards the tool is sufficient in order to start the next winding process. As the wire directly connects the coils, a wire separation of the finished coil is necessary before the extraction unit can take out the part. When using this technology, there is usually a larger wire waste compared to the linear winding technology. The remains can be extracted from the machine using pliers and a stroke.

The advantage of this setup is the limited number of motions in manufacturing, resulting in short cycle times. Benefits of the design are that the wire clamps are on the same axis as the winding tool, and that both the cycle times and distances on the rotary table are short. Terminating at the contact posts does not make sense for this technology, as terminating at the bobbin demands additional axes.

Application of flyer winding technology

Today, fully automated machines with multi-spindle flyer winding technology have established themselves and perform many manufacturing tasks. The benefits of this winding process become apparent especially for winding of stators with external grooves and rotors in combination with high production volumes. The advantages of this process mainly depend on the coil structure and the respective requirements of the winding. Stators, respectively rotors, have a special geometry, an off-centre installation of the coils seen from the stator rotational axis, and contain a number of different coils. Considering an evenly cylindrical bar, which is wound with wire, the bar can easily be set into rotational motion without adapting larger vibrations. The running smoothness of the rotating object on which the wire is placed is the prerequisite for a high-quality winding. Thus, flyer winding technology is used if a bobbin cannot be spanned on a spindle due to its outer geometry, or if the resulting imbalances are intolerable. Due to the flying wire, which revolves around bobbin in a relative distance, it is even possible to wind into very slim grooves with the use of two separate followers.

The flyer winding process is also suitable for bobbins with high masses. When considering the additional weight due to the wound wires, this becomes even more apparent. The winding itself can also cause imbalances in rotating parts. This leads to another condition, as it is only reasonable to process wires with wire gauges below 0.6 mm. This limitation is due to the necessary diversion of the wire in limited space, but with very high speed. With a significant limitation in process dynamics, wire gauges of up to 1.5 mm can be processed as well. However, the advantage of flyer winding is the high number of turns and small wire gauges, which can be produced at high winding speeds.

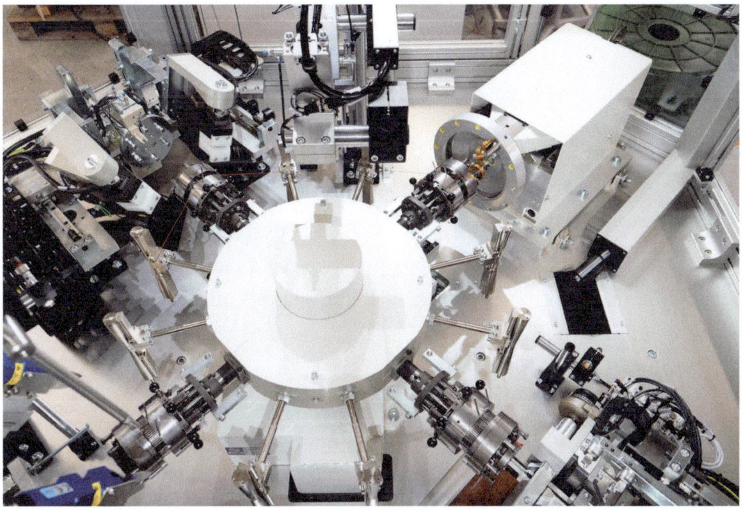

Follow-up process

Figure 3.49 Introduction to the follow-up process. Scan QR code to view video demonstration.

Flyer types

Due to the different applications that have arisen in the past years, two particular types emerged from a number of specific flyer types. Each can be characterised by its structure. They are the disk flyer and the bell flyer (Figure 3.50). For the disk flyer, the outer contour is similar to a disk. A pulley can be used for the wire guide and diversion, which feeds the wire more carefully. The main body of the bell flyer geometrically resembles a bell. Usually, wear-resistant guiding bushings are used to divert the enamelled copper wire, using highly alloyed or oxide ceramic materials. The main difference with respect to the application are the coil geometry and therefore the number of turns, as well as the pitch, the layer height and the coil length. A longer placement distance with higher wire tensile forces can be realised with the bell flyer, due to the shifted wire guide. The big mass of the flyer, which in turn has poor inertia and therefore lower dynamics for the manufacturing process, is a disadvantage. The bell flyer is usually equipped with nozzles, which allow high speeds of up to 12,000 rpm for smaller wire gauges. The disk flyer can be equipped with a nozzle wire guide or pulley wire guide, and is therefore suitable for higher wire gauges of up to 1.2 mm with winding speeds of up to 3,000 rpm. The two most important characteristics of the bell flyer and the disk flyer are their diameter and the bell depth.

Figure 3.50 Comparison of bell-, disk-, and template flyer
(Picture on the right: Otto Rist GmbH & Co. KG)

Another flyer type for stator manufacturing is an alteration of the disk flyer. Guiding jaws in the shape of a template are at the rotational center of the flyer. Therefore it is referred to as template flyer. An inner tool segment, which is navigated to the stator, completes the important tool sets. The guiding jaws are tools which are specific for each product or product line. The surfaces which are in contact with the wire are made from highly polished, corrosion-resistant material, in order to permanently reduce wire-friction. The one-sided mounting of the flyer arm, including the wire guide nozzle and the pulleys, causes an imbalance which creates larger bearing stresses at higher speeds. To counter this influence, the flyer must be balanced with all of its attached parts. Due to safety-related aspects, the balancing of the flyer head disks, bell flyer, or flyer disks, requires special care.

Spectrum of performance

A modular machine design enables flexible follow-up winding processes on the one hand, and linearly arranged multi-spindle machines on the other hand. For follow-up winding processes, the required modules are typically integrated into the cycling rotary table. Manufacturing speeds of up to 12,000 rpm on the flyer axis can be achieved, whereas the flyer diameter limits the feasible speeds. To divert the wire, disk flyers have guiding pulleys adjusted to the wire gauge. These are supported on axes with friction or roller bearings and therefore only allow up to 6,000 rpm. The final winding speed is following a speed curve which is adjusted to each winding task and consists of acceleration and a deceleration ramp. Both the inertia of the part and a uniform wire tension have to be considered for this. When designing the flyer, the arrangement of the diversions and wire exciting radii, which ultimately describes the revolution course of the wire, have to be adjusted to the winding task. The design of the diversion points determines the critical wire stresses. They limit the possible flyer speed and therefore determine the manufacturing speed of the whole process. The flyer winding process can also be used to manufacture orthocyclic windings, where the self-guiding properties of the wire on the coil surface are beneficial. The self-guiding properties allow a wire placement, where the residual stresses of the wire and the orientation on the winding ground are positively influenced. Additional guiding jaws can support the placement process. If a flyer machine is equipped with a CNC controlled placing gear, there are sufficient spatial degrees of freedom for a proper wire placement. In most cases, it consists of three independent servo motoric axes. For this arrangement, it is also possible to equip the winding head with multiple winding axes to enable a multiplication of the production. However, a follow-up winding process is significantly limited in such multiple flyer arrangements. This rational arrangement can often be found in the production of high volumes. Accordingly, the servo motoric drives for the placing gear axes have to be adjusted to the conditions and must be designed dynamically. This is because, while the winding head is very compact, various weights forces apply on the placing gear, which are created by the motoric axle drives including different numbers of winding spindles and flyer heads. A stable and vibration-reducing machine engineering is absolutely necessary, as almost all movements take place in the same direction. This application uses a nozzle flyer, as it enables terminating of coil or contact pins.

Fully automated winding process

Terminating on wire park pins or on contact posts can be designed similarly to linear winding technology. A multiple flyer arrangement is also possible, although this should be considered in terms of economic and technical aspects. Sub-assemblies similar to other winding technologies (placing gear) can be used. The structural arrangement of additional sub-assemblies, like wire park pin, active wire cutters or crimp pliers, depends on the processed product and can only rarely be standardised due to the different geometries. As the winding space fully surrounds the coil due to the rotating flyer, a stroke is necessary to feed all sub-assemblies. To fixate the wire, regardless of the pin type, the flyer head has to approach a certain position in both the direction of wire placing and angularly. Aside from the actual winding, often extensive wiring of the individual coils must be realised as part of the stator manufacturing, which leads to a complex termination at and around contact posts. These specific requirements have a significant influence on the length and angular position of the wire guide nozzle. Mandrels or levers can be brought into the winding space for additional loop placing. Similar to linear winding technology, flyer winding can be designed as a wire waste-free process, although the appropriate prerequisites, like clamping contacts on the products, have to be provided. As in other winding processes, active wire separation can also be realised with attached wire cutters or by tearing. Since a thin wire, which is characteristic for this process, is used in most cases, cost-effective tearing is preferred.

Fixture, counter bearing, tools

For follow-up winding processes on a rotary table, the bobbin fixtures have a similar or the same design as in linear winding technology. To wind the bobbin, it must be rotationally locked. This fixture and the flyer are designed specifically for each winding task. During winding, the stator must be centrically positioned to the flyer. Alternatively, in fully automated manufacturing it can be brought into and positioned in the winding space with a lifting station.

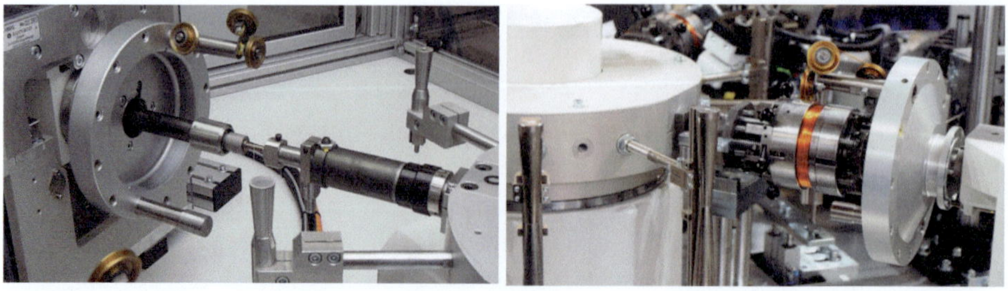

Figure 3.51 Design of flyer winding machines with and without counter bearing

As stators made from lamination stacks sometimes have a high dead weight, the dimensions of the fixtures must be chosen accordingly. The stator is seated in a position vertically to the orientation of its axis. A counter bearing is indispensable, because tensile forces act on both of the stator's front sides due to the rotational winding. An outward shifted bobbin can run smoothly only with a suitable counter bearing.

Additionally, a rotary fixture and therefore a switching mechanism is necessary for fully automated manufacturing, in order to cycle the stator from one tooth to another in one manufacturing routine. Further important parts of stator manufacturing are the guiding jaws, which are responsible for sliding the wire into the winding space, and must therefore be made specifically for a product or product series. The quality of the pick-ups and counter bearings is determined according to the same principles as in linear technology. Precise tools and a high positioning accuracy without balancing are necessary for excellent winding quality.

Figure 3.52 Types of guiding jaws

Wire course

The diversion of the wire, and therefore the guiding elements, have a large influence on the winding capabilities of a manufacturing machine. Similar to linear technology, the enamelled copper wire is led from a supply spool via a wire tension control system, which creates the appropriate wire tension. Then, the wire is fed directly through the drive spindle, via diversions and to the wire guide nozzle. The challenge lies in the design of the flyer. The wire has to reach the outer diameter of the flyer through the connection of drive shaft and flyer, where the wire guide nozzle, wire guide bushing or diversion pulley is located. In some cases, an additional non-rotating pick-up must be provided at the center of the flyer. This enables the fixation of the counter bearing or guiding jaw elements for the winding of parts with external grooves. At wire diversions in particular, special care is needed when choosing the guiding elements and the bearing positions, because the maximum speed of the flyer is very high. The construction is even more complex, if an additional placing stroke of the flyer arm is needed.

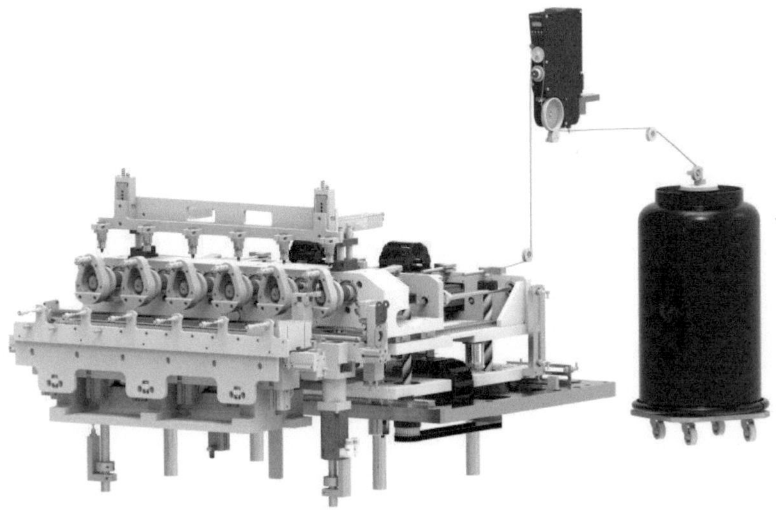

Figure 3.53 Wire path for a flyer winding machine

Special features in wire placement

Balancing elements for wire tension compensation, as well as guiding elements for wire guidance, are often necessary parts of flyer winding machines. The flyer winding deals with similar challenges as the linear winding. The wire placement on a circular cylinder coil can easily be evenly placed with a constant wire tension. However, flyer winding is also used for stator manufacturing. As the stator teeth are often designed to be long and slim, different proportions of edges, and therefore different placing wire lengths, result within one rotation of the flyer. However, they are not as severe as in other technologies, since the circular distance of the flyer wire guide must be designed in a way that ensures an acceptable sliding of the wire through the guiding jaws into the stator groove. If a suitable wire tension cannot be achieved, a balancing system should be provided after the wire tension control system (i.e., dancing lever system). For some wire gauge ranges there are combined devices which enable a safer process. However, the wire tension control design must be considered for each application and is a crucial step towards achieving high winding qualities.

Air coils

Flyer winding technology has a special application in the manufacturing of air coils. To manufacture stators with insert technology (Section 3.3.8), the required coils are premanufactured with the flyer winding process. Due to the rotationally pivoted tool, different coils with different numbers of turns are placed on individual fins. This special type of placing enables the manufacturing of distributed windings on the tool (Section 1.3.4). This particular flyer is referred to as template flyer.

Figure 3.54 Flyer winding machine with a 3-axis traverse system

Maintenance and cleaning

Due to the complex kinematics, fully automated flyer winding machines are equipped with numerous guiding elements. These require an appropriate maintenance and cleaning at certain intervals. A maintenance schedule provides information about the frequency and the kind of maintenance work necessary. Wear parts, and especially parts which are constantly in contact with the wire, must be renewed regularly. A clean wire is a prerequisite for high winding quality and good guiding properties. Bobbin or stator fixtures and guiding jaws should be manufactured and maintained in hardened and highly polished quality. Special attention should be paid to flyer disks or heads, as they are an important part of this technology due to their delicate wire guiding courses. Disk flyers with pulley wire guides are in focus because of their pulley bearing. This one is stressed by occurring satellite movements.

The flyer winding process is a very cost-effective process, as in many cases the number of NC-axes is smaller than in other processes. Typical applications are coils with high numbers of turns with random windings and armature winding. As there usually is no termination, the flyer winding technology is a winding process with comparatively short cycle times.

3.3.6 Needle winding technology

Introduction

Unlike the previous winding processes, the term needle winding technology is not derived from the type of wire placing, but rather from the geometric structure of the wire guide respectively the nozzle. The wire guide, which acts like a needle, directly navigates along the placement contour around the bobbin. Opposed to other winding processes, the needle winding technology has longer nozzles, which are adapted to the part which is to be wrapped (Figure 3.55).

Similar to flyer winding technology, the bobbin is typically tightly clamped. Needle winding systems, which clamp the bobbin but allow a rotational motion, are an exception. A rotary motion of the part is always necessary for wire placement. This is why there are pivoting needle carriers. The wire is not placed by a circular motion, but on a direct course along the winding geometry. The termination of individual turns, which depends on the product geometry, is usually manufactured fully automatically. This is why this winding process is very cost-effective and reliable even for high numbers of poles. The needle winding technology is mainly used for products with thicker wires and lower numbers of turns. Hence it is opposed to the flyer technology, where thin wires and high numbers of turns can be processed economically.

Example 1 Example 2 Example 3

Figure 3.55 Introduction to needle winding processes. Scan QR code to view video demonstration.

Applications

One distinguishes between three areas of application in needle winding technology. These application areas comprise a product category with externally grooved stators, internally grooved stators and pole chain winding, respectively. Needle winding technology is not suitable for traditional circular, quadratic or rectangular cylindrical single coils, as other technologies work much more economically for these types of coils.

Therefore, stator winding is of great importance in needle winding technology. Similar to flyer winding technology, all of the placement motions for this process are executed by the wire guide. However, the rotation does not take place via a circular motion. Instead, the wire is directly placed along the stator or the circumference of the tooth. The resulting dynamic requirements are a challenge for the manufacturing machines and, in comparison to other processes, limit the number of turns per minute.

Pole chains are multiple, linearly arranged and attached single-poles which are assembled to a stator after winding. Due to the existing connection, the single-poles just have to be arranged as a circular stator shape. An overview of the parts is shown in Figure 3.56.

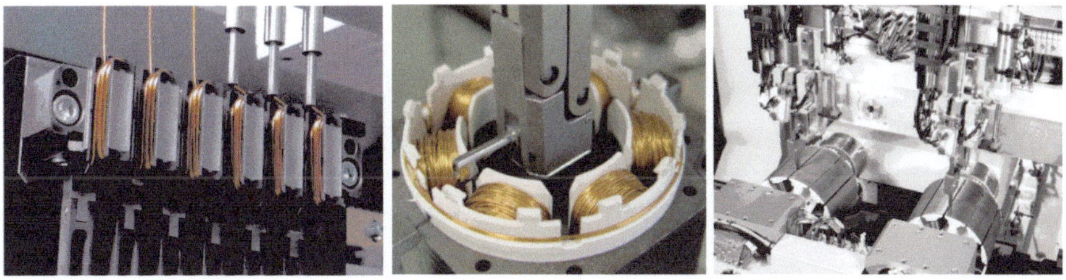

Pole chain **Inner-grooved part** **Outer-grooved part**

Figure 3.56 Areas of application for needle winding machines

Thicker wire gauges are processed in most applications using needle winding technology. This is why the needle carrier and wire guide nozzle are subject to special requirements. In needle winding technology, it becomes clear that winding technology is classified as joining by forming.

Machine types

Due to a variety of product categories, there are many different types of manufacturing machines in needle winding technology. Special types of needle winding have been established for the high demand of different motors with relatively thick wires and high fill factors. For the geometry of a stator it can be crucial whether the stator blank is picked up vertically or horizontally, and which dynamic targets can be achieved (Figure 3.57).

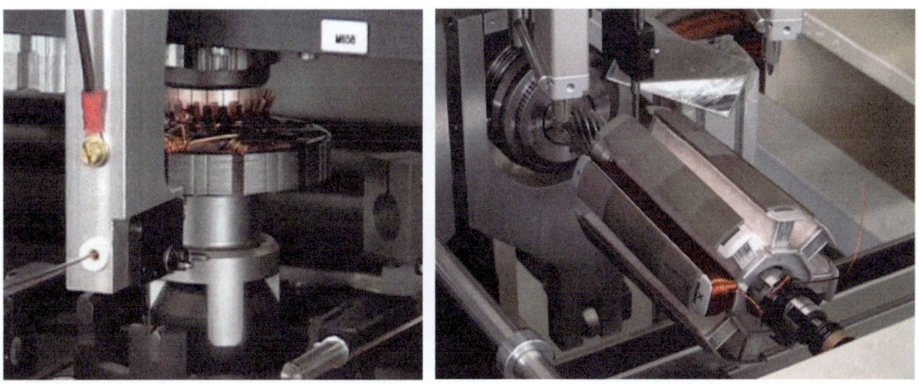

Figure 3.57 Design types - left: Vertical structure, right: Horizontal structure

The vertical arrangement of stators is often limited in that wires can horizontally be placed more evenly in very long grooves from a dynamic point of view. For very short grooves and therefore short placing strokes, a mechanical cam control or synchronized servo drives are used for high winding speeds. For chain winding, special needle winding machines are designed, which are adapted to each winding task in their systematic structure (Figure 3.58).

Figure 3.58 Pole chain winding machines

The basic setup of needle winding systems, including the placing gear, can be standardised in most cases. This has certain levels, depending on the number of simultaneously manufactured stators and their size. This shows that even in needle winding technology a duplication or multiplication of wire guides can be economically reasonable. The larger the processed wire gauge is, the higher are the forces that are applied, and the machine has to be designed with a higher stiffness. Due to the high utilization of manufacturing machines, they are often equipped with exchangeable parts or sub-assemblies, in order to wind different products. The limitations of stator manufacturing are lying in the groove width between the pole shoes, and therefore are defined by the distance between stator teeth or stator length.

Wire feeding and wire course to the wire feeder

Similar to the previous processes, the wire is pulled from a supply spool and fed to the wire guide via a wire tension control system. For winding of externally grooved stators, which are wound vertically, and chain winders, the wire supply is often located outside the manufacturing machine (Figure 3.59). This can prevent unnecessary diversions which would have a negative impact on the wire quality. The control systems are usually designed to a certain wire gauge range and must be adapted to each winding task.

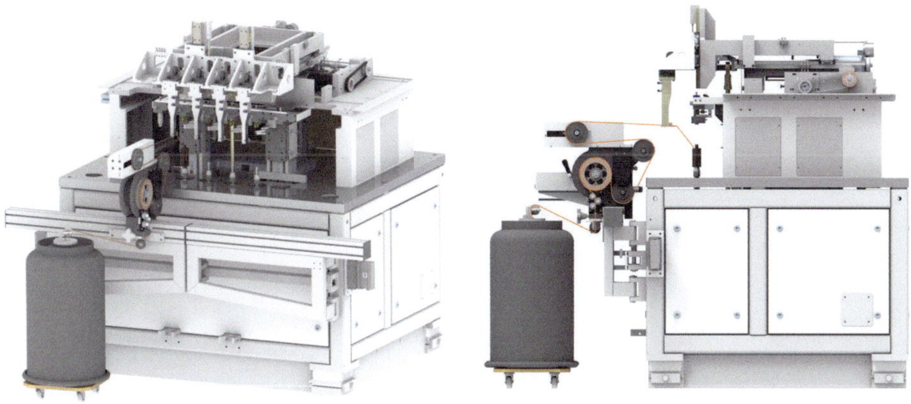

Figure 3.59 Wire path through the needle winding machine

The lubrication of enamelled copper wire has a very important task in needle winding technology. Since the friction forces which act on the wire are reduced with the lubricant, it should be carried along up to the wire's actual placement to protect both the tools and the machine.

The wire guide or needle carrier

In needle winding technology, there are special requirements for the wire guide. The needle is often designed to be longer than in other winding processes due to the handling deep inside the pole shoe. The wire guide nozzle is referred to as needle in this technology. The inner diameter is adapted to the wire guide. It has a highly polished surface and a significant influence on the winding quality. The wire often exits the needle at sharp angles and is at this point highly strained, just like the needle. To terminate and place the wire on the stator face side, the needle must be pivoted around an angle. This is especially challenging for internally grooved bobbins, as the stator determines the installation space for the wire guide. When winding tooth chains with linearly arranged pole shoes, three phases can be wound at the same time. Connecting by terminating after the winding process is unusual for pole chains, as this is done with other technologies after rounding the stator. As multiplication of the wire guide is possible in needle winding technology, multiple stators and phases can be wound at the same time. An adaptation of the other sub-assemblies of the machine, however, must be taken into account. For all needle winding processes the critical measure between the pole shoes must be considered. This also has an impact on the dimensions of the wire guide nozzle. Figure 3.60 shows the design of the needle carrier.

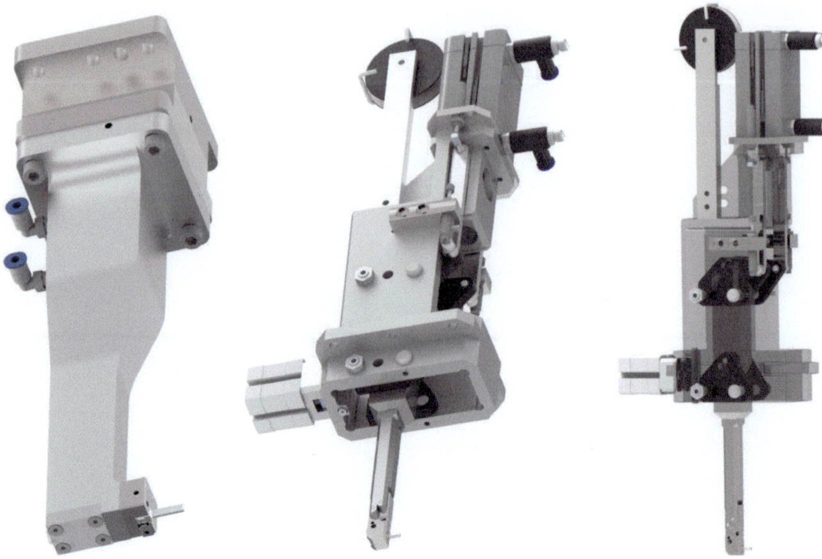

Figure 3.60 Needle carrier

Fixtures

When considering stator fixtures, different product categories can be distinguished. Fixtures for externally grooved stators grip inside the part. Due to the imbalance properties of the part, they are comparable to those of turning parts, but are not as critical because of lower speeds. Accordingly, internally grooved parts are gripped on the outer diameter, and are therefore constructed more elaborately in most cases.

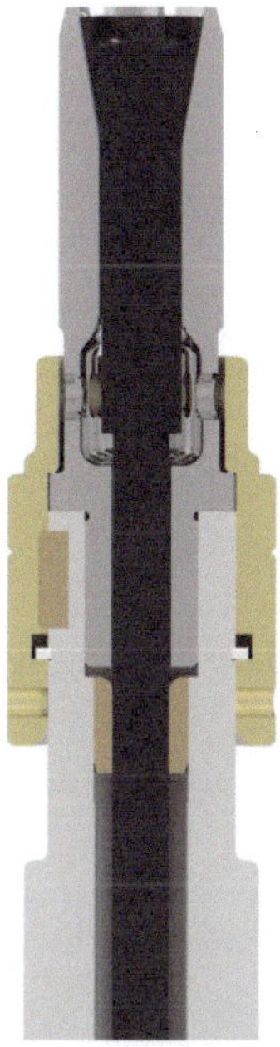

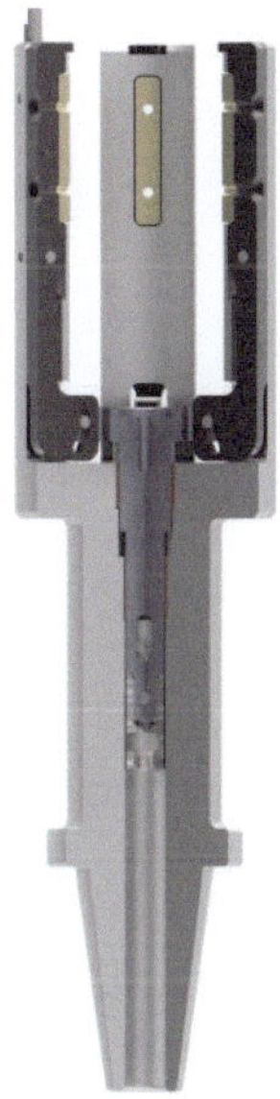

Fixture for outer-grooved parts **Fixture for inner-grooved parts**

Figure 3.61 Fixtures for needle winding

After each winding, the stators are switched from one pole shoe to the next via a switching gear or a CNC turning axis. A rotational fixture designed as force-fit or form-fit must generally be taken into account. Usually, product specific fixtures are intended for manufacturing single-teeth, respectively concentrated windings or stator chain windings. Initially, the internally grooved stators are aligned linearly before later being rounded. During the winding process, they are lined up like a chain and fixated on their circumferential surface, so that the pole shoe is accessible from all directions necessary for winding. It should be ensured that the parts which are to be wrapped are fixed well, so that the force applied by the placing is well absorbed by the tooth. To enable convenient loading and unloading, an easy clamping must be provided for both manual and fully automated processes. These special tools must also be of high manufacturing quality.

Fully automated winding process

For product manufacturing with needle winding technology, the wire is also fixated at a starting point. This can be an external wire park pin or a contacting post on the tool, or the stator. The difference to other winding processes is the processing of thicker wires. In this process, the loose wire end has a relatively stable shape due to its high stiffness, which does not necessarily demand clamping between the winding processes. Based on the motion sequence during placement, the switching gear axis can directly support the operating procedure and take over the rotational motion, as long as it is controlled by a servo motor. To prevent the needle from touching the stator, a precise synchronisation of the rotational motion and the stroke motion is necessary. Influencing factors for the maximum winding speed are, among others, the needle stroke, the stator rotational angle defined by the number of poles, the wire gauge, and the slot width. A special role is assigned to the pitch angle of a stator with pitched grooves. Both servo motoric rotational axes and a communication between two axes is necessary in order to ensure a proper motion profile. This also has an impact on the manufacturing times. Since the distribution of the stroke length to the rotational motion of the stator is significant, and the servo drives must constantly reverse in order to generate the reversed needle motion, a crank disk can be used. The initiation of the stroke motion is transferred to the crank disk via a rotational motion, which should contain a stroke adjustment, in order to achieve the benefits of a sinusoidal/semi-circular motion sequence. Depending on the design, up to 2,500 strokes per minute can be achieved. As the wire guide nozzle can be moved freely, the wire termination on the contact point can be performed with an additional pivoting of the nozzle. Similar to traditional linear winding technology, a contact pin or a lug is used to connect the individual poles with either a star or a delta connection. During the process, the combined weight of wire guide and needle carrier may lead to undesired vibrations due to the axis acceleration. This, in turn, can influence the winding quality negatively. Accordingly, alternative materials from steel are considered. By using the needle winding technology it is possible to manufacture whole sub-assemblies, including stator coils, wiring, and contacting on a single machine. As opposed to insert technology, and aside from the not fully utilized space between the poles, it is possible to wind motor coils on lower lamination stacks with smaller winding heads, which have a good fill factor.

Wire placement

Nearly every pole shoe has a rectangular or quadratic coils shape. Consequently, the winding speeds of this winding technique are constantly changing and the wire tension control is expected to balance the speed variations effectively. The acceleration and deceleration of the wire guide and the different forces on the stator therefore require firm tool fittings and stability of the individual axes. This is a pre-requirement for the manufacturing of high quality windings. The wire processing on small stators is usually based on concentrated windings around the individual pole shoes. Therefore, it is significantly different to the processing of windings with insert technology, for which the winding can be inserted over multiple teeth. The placement does not necessarily follow from one stator tooth to the other. Different winding topologies are possible as well.

The wire is bent after exiting the wire guide nozzle, since the wire guide nozzle moves past the placement contour sideways rather than with the front, like in other processes. This leads to a wire diversion of up to 90° and significantly strains the wire itself as well as the nozzle. It is often difficult to wind enamelled copper wires with larger gauges for small components with an orthocyclic scheme. The extreme bending of the wire before the placement is the reason for an undefined deformation geometry from the residual stresses. However, an ordered layer structure can be realised for large wire gauges with a precise manufacturing machine, using the properties of cylindrical bobbins. For better wire guidance, auxiliary tools may be used, as shown in Figure 3.62. Aside from guiding, they also pre-shape the wire. In accordance with the Bauschinger effect (Section 3.1.1), the wire can be bent in the opposite direction more easily.

Figure 3.62 Auxiliary tools for needle winding

A space the size of the nozzle diameter or the nozzle width has to be left between neighbouring poles, which is a disadvantage. The nozzle diameter or nozzle width is about three times the wire gauge. By using oval nozzles or nozzles with a trumpet-shaped opening, the relation of slot width and wire gauge can be reduced. The space between two neighbouring poles can therefore not be filled completely. An exception to this is guiding the wire outside of the slot, in order to utilise the remaining winding space. In this method, the precision of the wire guiding depends on the residual guiding properties of the wire. Currently, a special type of needle winding technology is being developed for the direct winding of distributed windings for larger stators. A new kinematics for more complex placing of the connecting wires is illustrated in Figure 3.63.

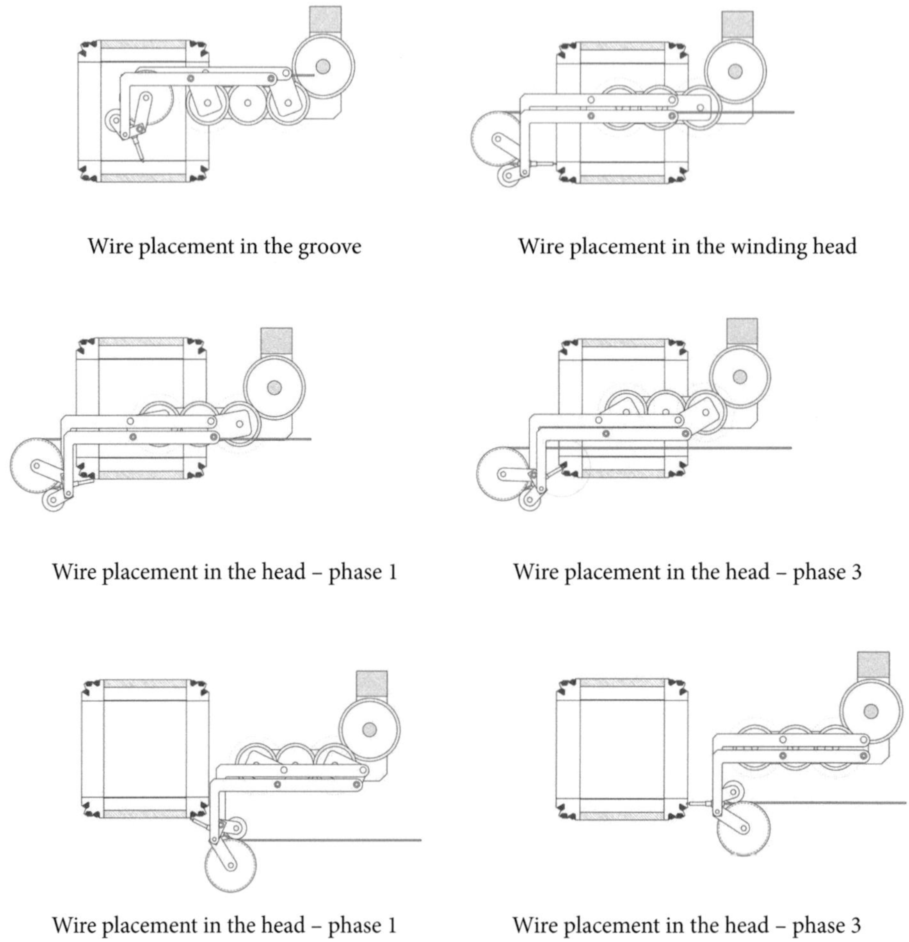

Figure 3.63 New developments for needle winding technology

Maintenance and cleaning

As fully automated needle winding machines contain a number of axes for the wire guiding motion, lubricating intervals, which result from maintenance plans, have to be carried out regularly on guiding elements and ball screws. Wear parts are those in touch with the wire, which have to be checked for their wear limits regularly as well. The wire guide nozzle is especially subject to strain.

The needle winding process is highly flexible due to the degrees of freedom for the wire guide. It is the only direct winding process which is suitable for winding internally grooved full lamination stacks and multi-phase pole chains in the area of motor manufacturing. The efforts to connect the windings can be reduced significantly. Externally grooved full lamination stacks are also suitable for needle winding technology when wound with a pitch or with large wire gauges.

3.3.7 Toroidal winding technology

As previously mentioned when discussing bobbins, the toroidal core winding technology is only used for circular toroids, which are usually closed. These ring cores are also referred to as toroid coils. The winding process can be divided into three different types. It can be executed by hand, wound with a hook, which is called hook winding, or the winding is created with a rotating wire storage. The typical wire path for toroidal winding with rotating wire storage is shown in Figure 3.64.

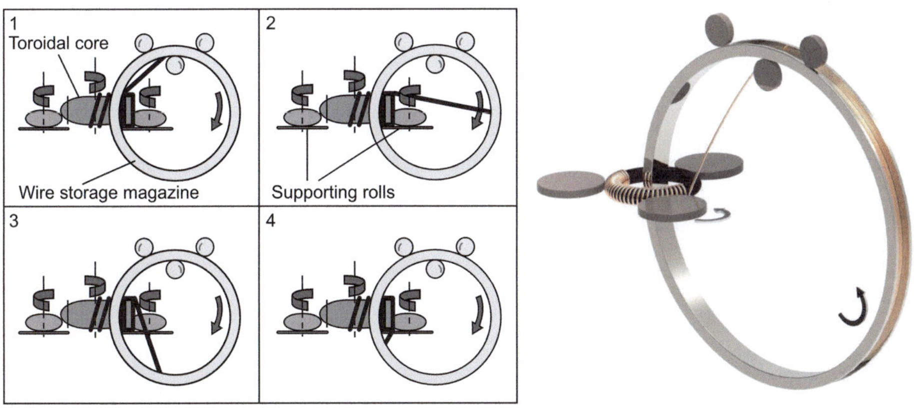

Figure 3.64 Wire path for a toroidal winding machine (according to [Fe-13])

Winding manually is characterised by guiding the wire by hand or with a tool. Loops or a magazine are often used as wire guidance for toroidal winding. The toroid can be wound as long as the tool can pass the core's inner diameter. In addition, precise wire guidance by hand enables the processing of small toroids with thin wires. However, the low level of automation and the high handling efforts lead to longer process times, which is why this process type is usually manufactured in low-wage countries.

Hook winding is a semi-automated process. An oscillating hook guides the wire through the center of the toroid. The wire guiding outside the toroid takes place manually through a machine operator. The toroid itself is fixed onto a rotatable three-point roller support. After terminating the wire on the bobbin, the wire end is passed onto the oscillating hook. With the use of a pneumatic cylinder, this hook is then fed through the hole of the toroid. After completing the turn assembly by hand, the initial position of the cylinder is approached and the wire is clamped into the hook again. This process is especially suitable for larger wire gauges. Due to the many repetitions of fixing the wire, the process speeds are low, so that this process is only suitable for smaller numbers of turns.

The most common type of toroid winding is the automated process using a rotating wire storage, which has to be filled before the actual winding process. As previously mentioned, clamping the toroid takes place in the three-point roller support while the wire storage is opened. Subsequently, the wire storage is closed and the core is wound. The winding process starts at the lower end of the U-shaped profile of the wire storage. It is performed by the rotation of the supply spool around the storage. By doing this, only the amount of wire required for one toroid is taken from the storage. The wire is then separated from the supply spool and is terminated on the toroid. The winding is created in the next step by the rotation of the magazine around the toroid. A pitched winding can be controlled via the relation of the wire storage rotation and the rotation of the toroid. This procedure and a picture of a machine is shown in Figure 3.65. The automated process achieves high process speeds and creates compact windings. Due to the single set up of the wire storage, it is suitable for applications with high numbers of turns. The necessity of inserting the magazine into the toroid hole is a disadvantage. As a result, the inner diameter of the toroid is limited. The handling of the start and end wires are manual processes, as these can often not be fixated on the toroid. Accordingly, full automation of toroid winding machines is limited.

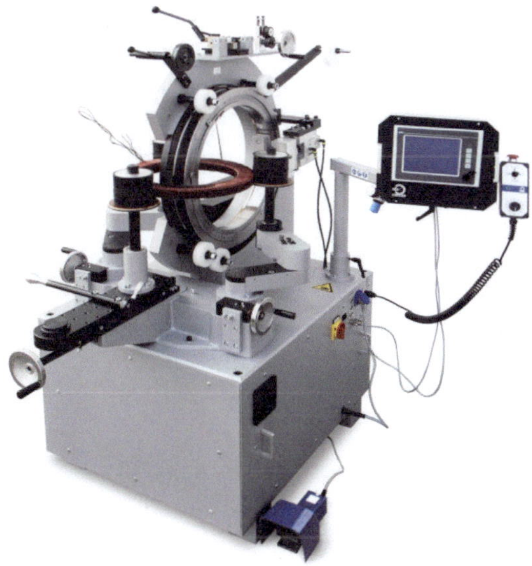

Toroidal winding

Figure 3.65 Demonstration of the toroidal winding process. Scan QR code to view video demonstration. (Picture: Ruff)

For winding toroid coils, there is no alternative to toroid winding. Especially for closed bobbins, this necessitates a rotation around the toroid. Although there are higher manufacturing efforts and manual processes, toroids are used because of their low magnetic flux leakage, lower iron losses and better power density. There are different levels of automation, and which to choose may depend on the location of manufacturing and the wage rate.

3.3.8 Insertion technology

For the inserting process, the required winding first needs to be manufactured as an air coil. This usually takes place in a prior flyer winding station, or a linear winding machine with a template, which winds the correct number of turns per groove in a transferring or inserting tool. A schematic overview of winding into the insertion tool is shown in Figure 3.66. Winding and inserting can either take place for each phase, where each phase is inserted separately, or all phases are successively wound onto the transferring or inserting tool and then inserted all together. The manufacturing of the coil must be executed without wire crossings per layer.

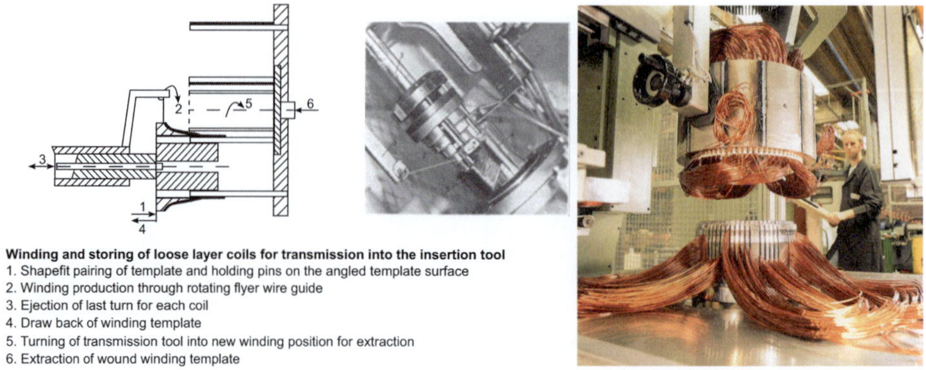

Figure 3.66 Winding preparation according to [Tzs-90]. Picture (right): Otto Rist GmbH

For the automated process, the turns are wound directly onto the inserting tool. Alternatively, they are applied to a transferring tool, in order to achieve a better orientation of the wires. Transferring the individual coils takes place without a feed, but by tilting the template. The last turns of each coil have to be transferred into the winding tool with an ejector. The insertion tool can carry a group of coils due its multi-staged structure. By rotating the revolver with the transferring or inserting tool, the next phases or coil groups of the winding can be created. This multi-stage structure enables complex winding topologies of distributed windings to be manufactured. It is therefore the only process, which allows a direct and automated assembly of these winding topologies for larger and closed full lamination stack with heights from around 200 mm. In the inserting process, the tool is first positioned below the stator and subsequently inserted. The tool consists of three components, the outer needle, the inner needle and the transmission tool in mushroom shape. The needles support the guiding of the wire into the stator grooves. During the inserting process, the transmission tool pushes the wound coils up inside the tool. Due to this inner motion, the coils are pulled into the stator grooves on the outside. The interaction of the components is illustrated in Figure 3.67.

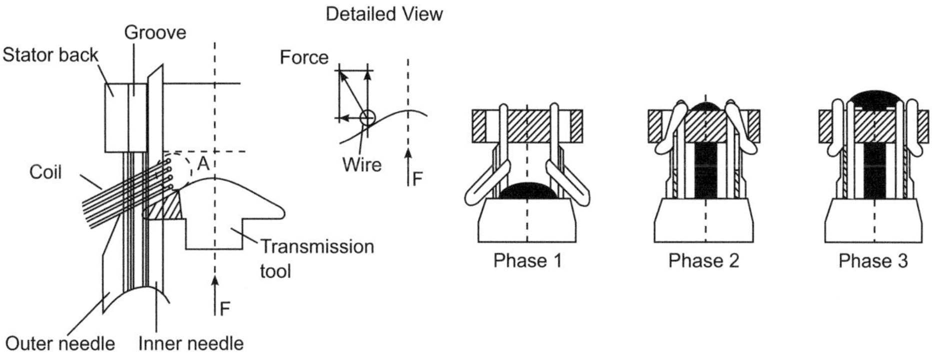

Figure 3.67 Process steps for inserting [Tzs-90]

This inserting process is carried out in three phases. First, the tool enters the stator. The transmission tool is then extended, initiating the insertion of the coil into the groove. Once the wires have entered the groove, a slot liner is added to prevent the wires from being pushed out of the groove after the inserting tool is removed. Finally, the coil is inserted. The term *inserting* originates from the operation area of the transmission tool. As it is located on the inside of the stator, the wires are pulled into the grooves using the wire tension applied from the inside. Aside from pushing the winding head, the transmission tool is extended so far into the groove that the winding is pushed against the groove ground. This is necessary to ensure sufficient space for the assembly of the slot liner. A structured surface of any machine components in contact with the slot liner enables a better guiding during the assembly. The process is critical at the point at which the wires are pulled into the groove. If an undesired arrangement of the wires results from inaccurate winding onto the tool or from bending the wire, the winding may jam. The coil wires are then pulled further by the feed of the transmission tool until the material malfunctions and the wire tears. These coils and possibly the stator become reject parts. The clamping factor was introduced to address this challenge. It describes the critical process window, based on the relation of slot width to wire gauge, and is illustrated in Figure 3.68. In this example the jam area is labelled with (2). If the groove width exceeds twice the wire gauge, two wires can slip into the groove simultaneously. This configuration is not critical and it is labelled with (1). If the groove is designed to be only slightly larger than the wire gauge, the wires slip into the groove individually during the inserting process. Clamping is unlikely in this case. This area is labelled with (3) and lies beneath the hatched line in the diagram. However, the mutual obstruction of two wires in the area in-between, labelled (2), may lead to clamping.

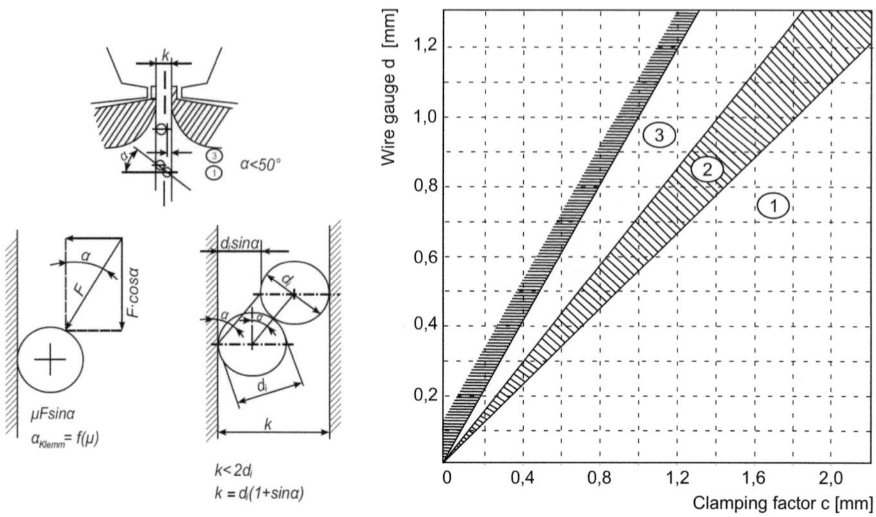

Figure 3.68 Introduction of clamping factor for the insertion process [Tzs-90]

If the groove is slimmer than the wire gauge, inserting is not possible at all. To prevent the possibility of clamping, inserting lamellas can be introduced which are attached to the transmission tool. Moving one of two flanking inserting lamellas along the inserting groove will break up any potential jams or blockages with additional forces in the insertion direction. Compared to alternative processes, axial inserting of pre-manufactured winding is the most productive stator winding manufacturing process, as illustrated in Figure 3.69.

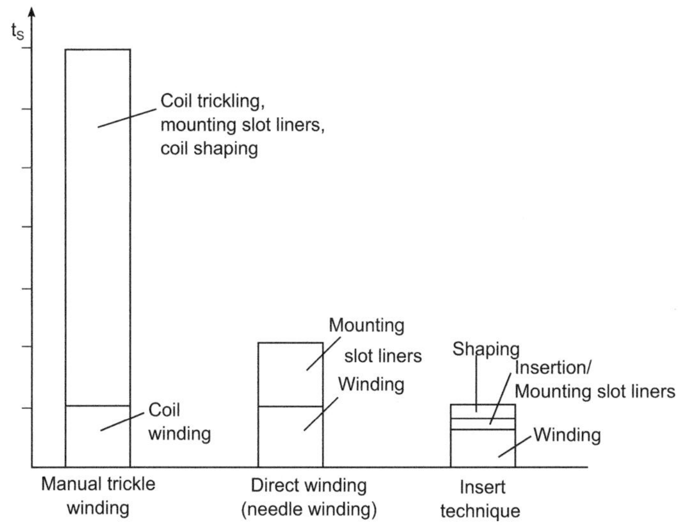

Inserting technology

Figure 3.69 Process times for stator coil manufacturing [Tzs-90]. Scan QR code to view demonstration video.

To compare processes, the process times of different steps of manufacturing are added. Depending on the hydraulic cylinders and the lamination stack length used, the process time for the individual inserting of a coil group or a whole winding can last 6 to 12 s. The slot liners are assembled between the winding and the groove in the direction of insertion using a stroke motion. Due to the previously assembled winding, the insertion of the slot liner is usually the most critical process step, as only very little space is left in the groove and the stiffness of the insulation material is low. Accordingly, the inserting process may malfunction if the slot liner is bent or completely deformed. This is often caused by the Hertzian stress, which is applied to the slot liner between the lamination stack and the winding. Once plastic deformation has taken place, assembly is no longer possible.

Aside from winding the layers without crossings, the cross-free arrangement of the wires in the inserting tool is achieved by using inserting needles. These are positioned in front of the stator tooth and guide the wires into the groove. The inserting needle length must be adapted to the wound stator lamination package. Additionally, the stack height of the wound layer has to be considered.

The stack height is also a criterion when estimating the process capabilities for the product. The winding task determines the design of the needles. Important parameters are the distance between the needles, and the profile shape, or stiffness.

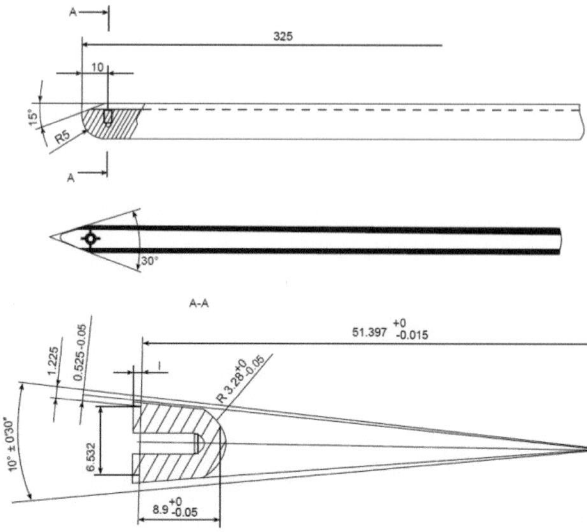

Figure 3.70 Typical parameters of a needle geometry for insertion tools according to [Tzs-90]

Tab. 3.4 Characteristic parameters for insertion equipment

	Insertion machine	Insertion center	Insertion equipment
Inner stator diameter in mm	12 … 400	55 ...250	
Outer stator diameter in mm			360 … 1000
Stator lamination stack length in mm	20 … 1600	15 … 250	240 … 1000
Number of grooves	12 … 72	12 … 72	24 … 48
Wire gauge in mm	0,1 … 2,0	0,12 … 1,4	0,12 … 1,6
Number of parallel wires template winding	2 … 20	2 … 20	12 … 30
Number of parallel wires flyer winding	12	12	-
Speed of template winding per minute:	300 … 1400	-	100 … 300
Speed of flyer winding per minute	2100 … 4000	3000	-
Coils/ groups	3… 8	3…8	3…4
Coil size in mm: winding length	1000		
Coil size in mm: Coil envelope	120 … 330	800	198 … 3300
Insertion time/operation	8 s … 90 s	6 s … 60 s	5 … 12 min
Insertion force in kN	25	5	3… 15

The inserting process is the most commonly used process to manufacture distributed windings in closed stators. It has a market penetration of around 80 %. This is due to the short cycle times and the broad field of possible applications in terms of stator and winding geometry. By using a two-step process with winding on an inserting tool and the actual inserting of the winding, only random winding schematics can be created. Compared with other winding processes, larger winding heads result. This is why manual trickle winding processes or direct winding processes are used for special applications with higher requirements.

3.3.9 Trickle winding or Fed-in winding

The trickle winding is the oldest winding process. Today it is mainly used for manual repair of electrical machines, as well as in the manufacturing of high performance machines. The trickle winding is characterised by the highest winding quality for electric motors, as the position of every individual wire in the groove is manually adjusted and optimised.

The process begins with the insulation of the stator lamination stack. Depending on the winding topology in the shape of a single-layer or double-layer winding, this requires a different number of insulation papers (Section 1.3.6). The insulation papers are laid down and folded. Subsequently, they are axially pushed into the slots. This process can be performed either manually or automatically. Folding the papers in the corners and on the edges of the slot prevents possible damage to the wires. A typical stator structure as well as a possible slot cross section for a double-layer winding is shown in Figure 3.71. The insulation paper juts out of the stator by around 1 to 1.5 cm to ensure the insulation function. The insulation thickness varies between 0.1 and 0.65 mm and depends on the insulation material which is used. This is defined by the dielectric strength of the motor. A lower strength is desired in order to increase the electrical fill factor. Typical insulation materials include polyester, polypropylene or polyamide. Typical products include Nomex, Mica and Dacron-Mylaer-Dacron (DMD).

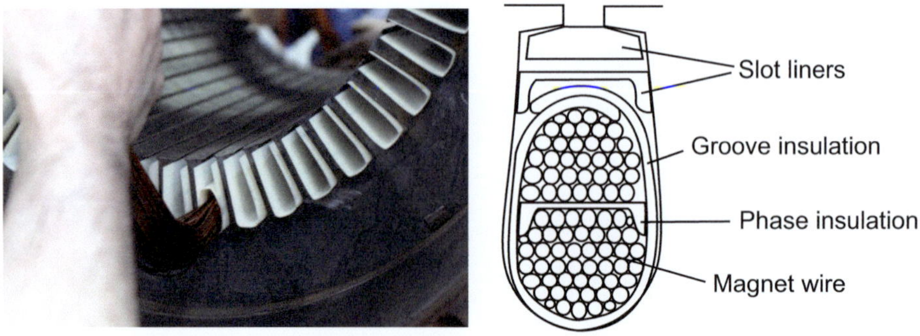

Figure 3.71 Insulated stator and groove insulation structure [Toliyat 2004]

As in inserting technology, the coil is wound outside the stator. Large coils are typically wound semi-automatically. Medium-sized and small coils are wound on fully automated winding machines. The example shown in Figure 3.72 shows the semi-automatic manufacturing of a winding with twelve parallel wires which are placed on the template by hand. The distance between the template ends defines the required coil size. The winding begins by knotting the strand to the template. The winding is created with a mechanical winding and manual guiding. For transportation, the wound strands are fixated with a polymeric tape or a cable tie.

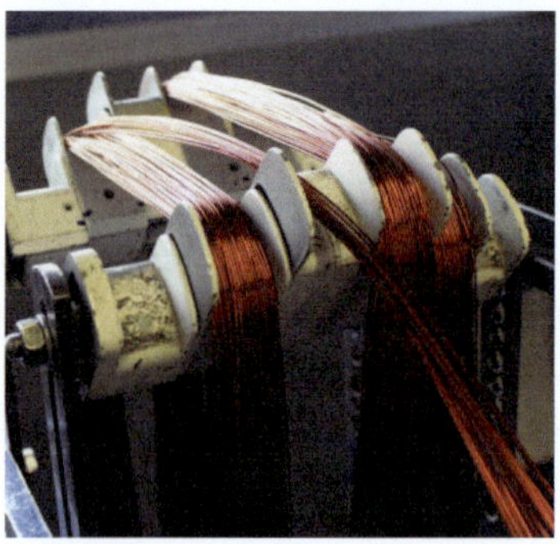

Figure 3.72 Manufacturing a coil strand

The coils are carefully removed from the winding template and prepared for insertion into the stator. The stator is already insulated. Now the actual characteristic process step follows the trickling of the winding. In the process depicted, the trickling is performed by two workers due to the large stack length. One person can handle smaller geometries. As with the inserting process, the process speed can be determined from the slot width and the stator length. The aim of the trickling is to place the wound coil strand into the groove. To prevent wire damage at the slot edges, the part of the coil that is not trickled into the stator is protected by insulation paper. The trickling itself describes a twisting finger movement of a coil strand which is held in the worker's hand. This is achieved by placing individual wire bundles between thumb, index and middle finger, and separating them into individual wires by rubbing in opposite directions. In case of larger bundles, the coil strand can be separated into multiple smaller wire bundles. A smaller number of conductors or a smaller assembly volume enables an easier assembly for the worker. For this, the lower coil strand is inserted from the front of the stator over a certain section. Feeding the remaining wires is now easier due to the previous separation via trickling, and they follow the part of the coil that is already placed in the groove due to the wire residual stresses. The remaining unplaced wires outside the slot are tightened and subsequently separated using the trickling motion on the other side of the coil. This leads to the insertion of further wires into the slot. This process sequence for manual insertion without tools is illustrated in Figure 3.73.

Figure 3.73 Process chain for feeding in: positioning and spreading the winding, insertion of single wire with tightening and subsequently feeding new wires

This process is repeated until the whole coil is placed in the stator slot. Alternatively, auxiliary tools can be used for the insertion. As an example, a wooden wedge can be used to push in the remaining coil strand once part of it has already been assembled. The wooden wedge prevents the winding from falling out of the slot and compacts it at the same time. It is important to ensure the wires are free of any crossings when inserted. If crossings develop, the strand must be removed from the groove and re-assembled. The trickle process sequence using a tool is illustrated in Figure 3.74.

Figure 3.74 Usage of auxiliary tools for winding feeding: wooden wedge

The wooden wedge used for various handling steps is a slim, flat, long wooden stick. It is placed in the slot opening at the beginning of the stator and is pushed onto the wires. Then it is pushed all the way through the slot until the stator end. In doing so, the outside wire parts are pulled into the slot. This process also enables compacting of the winding, as illustrated in Figure 3.75.

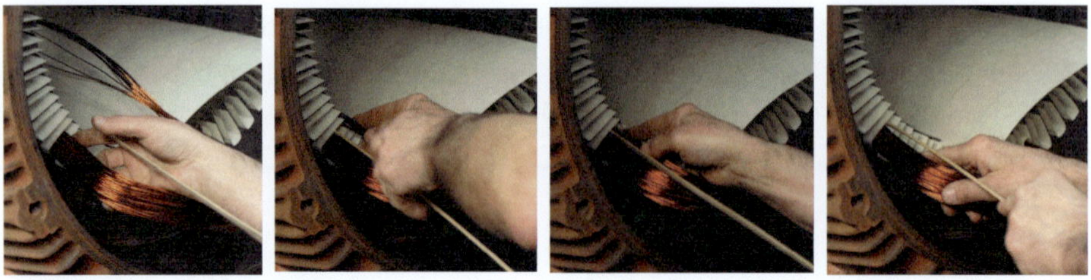

Figure 3.75 Feeding of outer strands

After inserting the wires into the slot, the winding head can be formed by using a wedge. For compacting, it is important to apply both tension and pressure on the wires in direction of the outer radius of the strand, if possible near the groove ground. This is necessary to create space for further coils. The winding head must have a defined shape to accommodate space for the complete winding head. Processing the winding head after multiple coils have already been assembled is quite difficult and may lead to wire damage. The forming is illustrated in Figure 3.76.

Figure 3.76 Forming of the winding head

Due to the long tradition of trickle winding technology as a training element for electrical engineering, numerous auxiliary tools have been developed. These include the folding stick or the trickle needle. The folding stick is a flat, wedge-shaped tool, rounded at the end, which is inserted into the slot on the upper side of the stator, and compacts the winding with its pressure. The trickle needle is a fine instrument to place and move individual wires in the groove. It is used when wires cross or if they form an unsuitable arrangement.

After assembling the remaining winding, and forming the winding head, the slot liner is inserted axially by hand or with pliers. Optionally, phase insulation can be assembled in the winding head. Contacting of the different coils to the desired winding topology is done manually. The winding is adjusted by a secondary insulation, e.g. a resin potting (Section 3.4.6).

Due to the manual characteristics, the trickle winding demands long process times and is therefore only suitable for small series and motors with high requirements for the winding. Typical durations for manufacturing distributed stator windings are 5 to 12 hours. However, due to the manual assembly of each individual winding wire it enables the best electrical and mechanical properties. Difficult winding topologies are manufactured with trickle winding to ensure the necessary space for the coil contacting, while maintaining a small winding head.

3.3.10 Selection methodology for winding processes

The appropriate winding process has to be chosen without considering any following processes, such as bobbin assembly or other steps. A suggested procedure is described in this section. The winding processes have been explained in detail in the previous sections and are therefore only briefly discussed in terms of typical applications.

The linear winding technology is suitable for three types of applications:

- Small rotationally symmetrical parts with high numbers of turns and small wire gauges (0.5 mm), due to the high productivity
- Rotationally symmetrical parts with large wire gauges (> 2 mm), due to the stress limitations of the machine elements resulting from the wire gauge
- Parts with profile wires, as their placing requires a simple wire guidance in order to prevent damaging of the wire insulation.

The flyer winding technology is also suitable for three types of applications:

- Parts which are not rotationally symmetrical and have high numbers of turns and small wire gauges (< 0.5 mm). These are, due to their geometry, not suitable for linear winding.
- Heavy parts with high numbers of turns and small wire gauges (< 0.5 mm), because they are not suitable for linear winding technology due to their weight, and the high number of turns would result in long process times for needle winding.

- Traditional armature winding technology, as the efforts to create a distributed winding can be reduced by turning the rotor in the clamping device. With the use of double flyers, two wires can be wound in parallel, which enables a symmetrical winding structure. This is achieved by winding two coils on two bobbins or four slots simultaneously.

For winding multiple wires, double flyers can be used for certain parts in flyer winding technology. As with other winding processes, bifilar winding with two-wire technology can be used as well. However, as opposed to the second technology, the first technology uses two separate wire guides.

The needle winding technology is suitable for four groups of applications:

- Externally grooved parts, like full lamination stack rotors, with a small slot width and high wire gauges, as an outside wire guidance is possible
- Parts with high masses and difficult slot access or access to the winding space
- Parts including placement of contacting wires, which makes this process interesting for pole chains with multiple phases as well as parts with higher numbers of contacting wires.

The inserting technology is used for internally grooved parts like full lamination stack stators with a distributed winding, such as a wave winding, and when manufacturing requires short process times and low production costs as well as medium product requirements. The manufacturing of concentrated windings is possible but rarely used.

The inserting technology is suitable for two groups of applications:

- Internally grooved parts, like full lamination stack stators with a distributed winding
- Externally grooved parts, like full lamination stack stators with a distributed winding.

The toroid winding technology focuses on the toroid coil, meaning round and closed parts. The toroid winding technology is therefore suitable for one group of application:

- Closed part, like toroids with a concentrated or distributed winding

The trickle winding technology is used for full lamination stack stators which have a high fill factor or a complex winding topology. It is used in the production of motors with the highest requirements for power density, but also for repairs or large machines.

The trickle winding technology is suitable for four groups of parts:

- Internally grooved parts, like full lamination stack stators with a distributed or concentrated winding
- Externally grooved parts, like full lamination stack stators with a distributed or concentrated winding
- Parts with large dimensions, for example in diameter or stack length
- Parts with high weights, such as large motors

To choose the right winding process, the part requirements are compared and evaluated with the process properties as shown in Figure 3.77.

Evaluation according to VDI 2225 0 = dissatisfactory 1 = sufficient 2 = satisfactory 3 = good 4 = excellent / ideal		Comparison in pairs according to Pahl-Beitz		Needle Winding pole chain all phases	Needle-winding stator inner-grooved	Needle-winding stator outer-grooved	Flyer-winding stator single poles	Flyer-winding stator outer-grooved	Flyer-winding pole chain of one phase	Linear-winding stator outer-grooved	Linear-winding stator single-poles	Linear-winding stator pole chain of one phase	Needle-winding stator inner-grooved (Hook-winding)
		Priority	Weighting										
1	Wire quality after winding	3.00	0.03	2	2	2	3	3	3	4	4	4	1
2	Electric fill factor	1.00	0.10	4	2	4	3	2	3	2	4	2	3
3	Thermal properties of winding	2.00	0.07	3	2	3	2	3	3	2	4	2	2
4	Ability for automation	2.00	0.06	4	4	4	4	4	4	4	4	4	4
5	Pre- and post-processing of stator	3.00	0.01	3	4	3	2	3	3	3	2	3	4
6	Phase insulation	3.00	0.02	4	3	4	3	2	3	2	4	3	0
7	Winding symmetry	1.00	0.08	4	2	4	3	1	2	1	4	3	2
8	Stator symmetry	1.00	0.08	3	4	4	3	4	3	4	3	3	4
9	Production volume winding machine	3.00	0.02	2	1	1	3	3	3	3	3	3	1
10	Investment stator production	3.00	0.03	2	4	4	1	3	2	3	2	2	3
11	Number of contacts	1.00	0.08	3	4	4	2	3	3	3	2	3	4
12	Electrical connectivity	3.00	0.01	2	4	3	2	3	2	3	3	2	4
13	Flexibility regarding connectivity	2.00	0.06	4	3	3	4	3	4	3	4	4	3
14	NVH	2.00	0.06	3	2	4	3	1	3	3	3	3	2
15	Variety of products, setting efforts	1.00	0.11	4	4	4	3	3	3	3	4	3	3
16	Number of stator parts	2.00	0.06	3	4	2	1	2	3	2	1	3	4
17	Windability of large wire gauges	2.00	0.07	2	2	2	1	1	1	4	4	4	2
18	Defined wire guidance for orthocyclic winding	2.00	0.07	4	3	3	4	2	2	2	4	3	2
Total		37	1	56	54	58	47	46	50	51	59	54	48
Quality rating				0.83	0.75	0.85	0.68	0.62	0.70	0.69	0.86	0.75	0.70

Figure 3.77 Comparison and methodology for winding process choice

The methodology used is the paired comparison approach developed by Pahl-Beitz. This approach describes the evaluation of different solutions. The evaluation can take place in multiple phases, and is performed according to the part criteria. In order to work out the relevance of the criteria, they have to be evaluated with respect to the product. The solutions are always specific to the product and can consist of a combination of a specific product design and the corresponding winding process. Therefore, manufacturing of a stator winding is possible with the following combinations:

- Needle winding of a pole chain with subsequent assembly
- Needle winding of an internally grooved stator
- Needle winding of an externally grooved stator

- Flyer winding of single-teeth with subsequent assembly
- Flyer winding of an externally grooved stator
- Linear winding of single-teeth with subsequent assembly
- Inserting for an internally grooved stator or other geometries
- Trickle winding for an internally grooved stator or other geometries

These variants require different consecutive processes and have different electrical and mechanical properties depending on the bobbin design and the winding process. Accordingly, the evaluation of the suitable process is important in terms of simultaneous engineering, i.e., the parallel development of a product and the production technology.

The most suitable winding process for a particular product is determined by the given technical and economic constraints. Aside from technical constraints, different solutions can be suitable for specific countries, depending on e.g. the local wage level.

3.3.11 Analysis of winding faults

The analysis of winding faults in commissioning or series production requires expertise. The challenge of analysing is the large variety of errors, as every winding product has its peculiarities and possible sources of errors due to its specific design. The purpose of this section is therefore to raise awareness for possible sources of errors and to provide the methodology for fault analysis. First, typical winding faults are characterised.

Obvious defects are errors in the layer structure of the coil. As illustrated in Figure 3.78, a gap, a crossing of wires, a double winding due to slipping of a layer or a faulty or missing flange winding, as well as a faulty layer step can be responsible for the creation of a random winding instead of a layer winding. Unclamping of the winding fixture, due to the resulting winding pressure, can be detected by an alteration of the coil cross-over direction. The consequence would be a higher electrical resistance even with the correct layer structure.

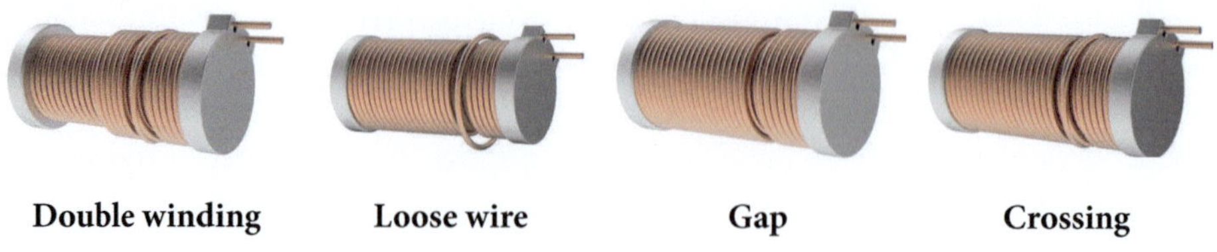

Figure 3.78 Selection of typical layering defects

Another defect that is easy to detect is the deviation of the winding geometry. For instance, it can be conical, concave, or bulgy. Typical for this error is a reversing on the coil flange that is too slow. Deviations of the bobbin dimensions, such as a smaller inner diameter or deformations, as well as faulty winding dimensions can occur. Several typical contour defects are illustrated in Figure 3.79. Stresses in the winding process are usually visible as wire damages or wire tears, and are therefore easy to identify. Aside from mechanical defects, a wrong coil resistance or inductance, or lacking of high-voltage strength are possible. A faulty termination or a faulty bonding of the coil are further error patterns.

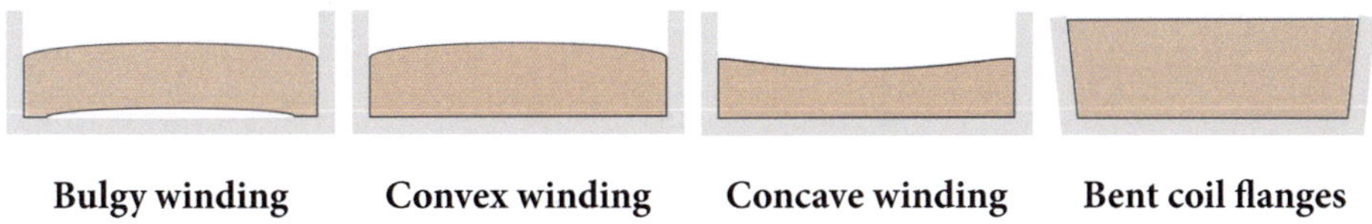

Figure 3.79 Selection of typical geometric errors

For the analysis of an existing defect, the different aspects of the winding process have to be considered. These are listed by categories in Figure 3.80: human, machine, method (process), material, environment and measurement. If a defect occurs in an otherwise stable series production, it is advisable to analyse current process changes, the supplied semi-finished products or the conditions of the production environment. If there are no obvious error sources or changes to the production conditions, an analysis of the influencing factors must be performed.

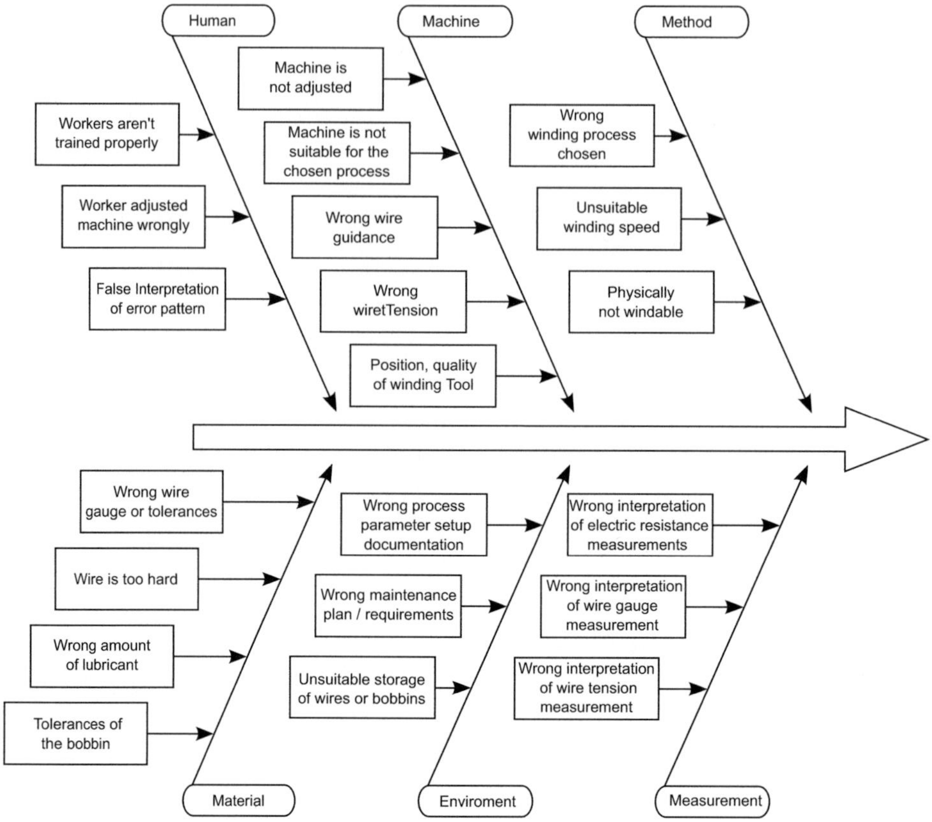

Figure 3.80 Influencing factors for winding technology

Due to the properties of the process chain, these influencing factors can become extensive. In order to provide some indication for possible sources of errors, Figure 3.81 shows influencing parameters for the wire, bobbin and process, and their interaction with typical defects. If this initial analysis is insufficient, a detailed analysis of the individual context will have to be performed. As an example, a fault tree analysis was performed for a faulty electrical coil resistance, as illustrated in Figure 3.81. Each parameter is analysed systematically. As in logical connections, individual parameters (OR connection), or multiple dependent parameters (AND connection) can be responsible for the defect.

	Interaction table for winding errors 0 = No influence 1 = Weak influence 2 = Distinct influence 3 = Strong influence	Typical Deviation range (0-5)	Resistance wrong	Turn to turn, or turn to body fault	Faulty dielectric strength	Wire damage	Wire tear	Fault in layer structure (random winding)	Loose turn	Bulgy winding	Conical winding
Process	Spindle speed ramp up	0	1	1	1	1	1	2	3	0	2
	Winding speed	1	3	3	3	3	3	3	3	3	3
	Machine stiffness / damping	4	1	1	1	1	1	2	2	0	1
	Castor angle (WG)	2	1	1	1	1	1	3	3	1	1
	Feed rate (WG)	2	1	2	2	1	1	3	1	1	3
	Turning point (WG	3	1	2	2	1	1	3	3	1	1
	Exit angle out of WG	0	3	3	3	3	3	3	3	3	2
	Wire tension	5	3	3	3	3	3	3	3	3	2
	Wire oscillation (Betw. Bob & WG)	3	1	2	2	2	2	2	2	3	2
	Free wire length (Betw. Bob & WG)	1	2	3	3	3	3	3	3	3	1
Wire	Amount of lubricant	2	3	3	3	3	3	3	1	1	1
	Wire gauge	2	3	1	1	1	1	3	3	1	1
	Insulation thickness	1	1	3	3	3	3	1	1	0	0
	Wire hardness	1	3	3	3	3	3	3	3	3	3
	Max. elongation at break	1	3	3	3	3	3	3	3	3	3
	Tensile strength	1	3	3	3	3	3	3	3	3	3
	Spring-back behavior	2	3	3	3	3	3	3	3	3	3
Bobbin	Structure winding ground	3	0	3	3	3	3	3	1	0	0
	Aspect ratio	3	3	3	3	3	3	3	1	2	2
	Winding width	3	0	1	1	3	3	3	3	1	1
	Bending radii	3	3	3	3	3	3	3	2	3	3
	Bobbin stiffness	4	1	2	2	1	1	3	1	2	2
	Friction behavior	1	0	1	1	1	1	1	3	2	2

WG - wire guide Bob - bobbin

Figure 3.81 Classification of typical errors and influencing parameters

Step by step, the interdependency of even minor parameters can be checked. As an example, a faulty wire length, which may result from a faulty bonding process, could create a winding short circuit that reduces the coil resistance.

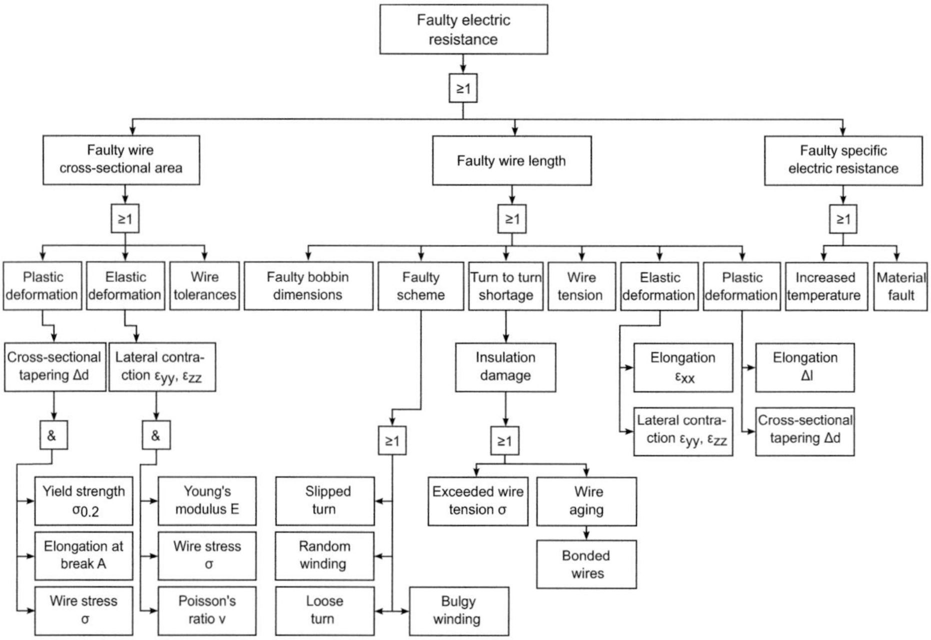

Figure 3.82 Development of a failure tree analysis (FTA)

Based on the FTA, product specific tables for possible causes of winding defects can be created. The development of critical product properties is performed by the manufacturer in a failure modes and effects analysis (FMEA) and is suitable for the basic elimination of process errors.

Analysing winding defects is complex, as they can usually not be directly associated to their sources. In an established process, variations in the quality of semi-finished products (bobbin, wire) are often the trigger. Initially, during the development of a routine for a new winding task, the different sources of error must be ruled out step by step in order to localise the origin of the winding defect. This requires a lot of expertise from the winding machine operator or the process developer.

3.4 Peripheral processes

Figure 3.83 Example for welding processes

Depending on the application area and the coil design, different upstream and downstream processes are used. From the pre-assembly to the winding readiness, the contacting process for the start and end wire, as well as various electrical, mechanical and optical testing processes often belong to the process chain. Bonding processes are only used for air coils. The coil test is performed manually or automatically depending on the degree of automation. The assembly of further elements can be found multiple times depending on the functional location. The secondary insulation most often takes place in motor windings, but also in thermal or electrically stressed coils. These processes are explained in more detail in the following section.

Accordingly, this section is aimed at readers from manufacturing or design of winding goods.

3.4.1 Preparing assembly processes

As discussed in Section 1.2.4, many bobbins require auxiliary elements for wire guiding, like pins, or a separate insulation, before they can be wound. These elements can be looked up in this section. Accordingly, facilities for the production of winding goods demand upstream assembly chains for these elements. Typical processes for the assembly include pressing-in of elements or adhesive bonding. An example, where two pins and an anchor plate have to be assembled, is illustrated in Figure 3.84.

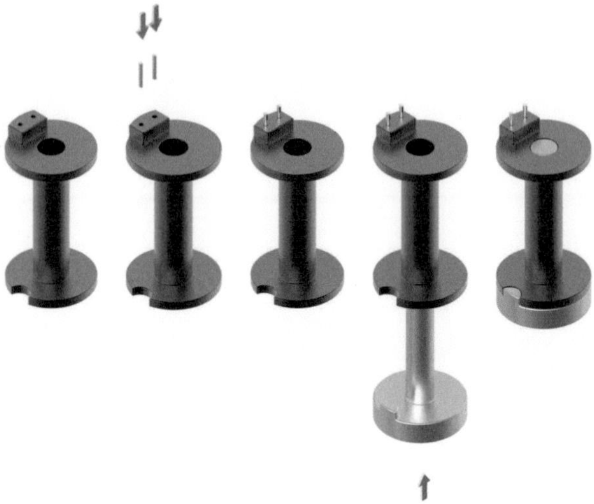

Figure 3.84 Assembly steps for coil bobbins

Figure 3.85 describes the stator assembly sequence for a full lamination stack. It requires groove insulation and an end disk to guide the connecting wires outside the groove prior to winding.

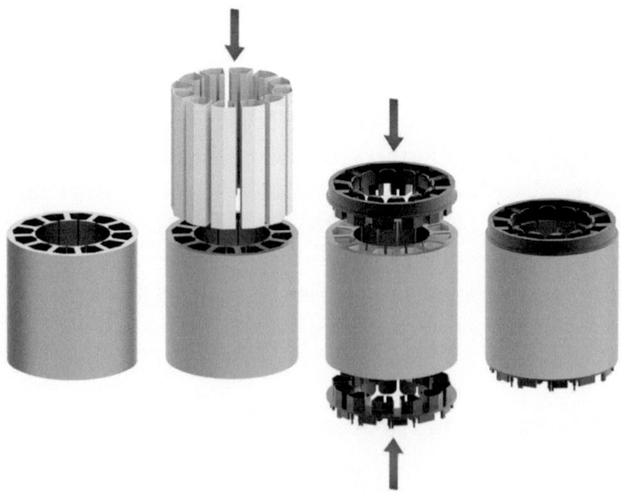

Figure 3.85 Assembly steps for full lamination stacks

For practical purposes of automation, the winding readiness is checked after the assembly of different elements. The visual inspection checks geometry features which are critical for the winding process. The test also ensures that only suitable parts are supplied for the winding process.

3.4.2 Contacting processes

The choice of suitable contacting processes is primarily determined by the wire properties. For a thermal contacting, the operating temperature of the insulation must be considerably exceeded. This has already been introduced for different systems in Section 2.3.2 in Table 2.10. If the wire was designed for high temperature applications, skinning can be done to contact the wire by soldering or bonding, for example. Due to the number of existing contacting processes, a basic introduction of the functionality and the machine structure for the individual processes is dispensed with at this point. In fact, procedural constraints are shown in terms of the winding technology in Table 3.5. The listed process limitations are empirical values. It is therefore possible to overcome some process limitations in certain process configurations.

Tab. 3.5 Characterisation of technical limits for contacting processes

	Process limit
Soldering	– Temperature stability of the insulation (temperature optimised wires) above 300 °C/350 °C need to be stripped – Lead free is limitation
Arc welding	– Temperature stability of the insulation (temperature optimised wires) need to be stripped
Insulation-displacement contacts	– Current limit of 10 A
Arc or plasma welding	– Limited material matching: CuSn1,5 & Cu not possible
Resistance welding	– Limited material matching: Aluminium not possible
Ultrasonic welding	– Limited material matching
Laser welding	– Limited material matching – Problems with contamination of insulation
Thermal compression welding	– Max. wire gauge up to 0.8 mm
Wire bonding	– Temperature stability of the insulation (temperature optimised wires) – Max. wire gauge up to 0.8 mm
Crimping	– Min. wire gauge of at least 0.1 mm

Generally, aluminium wire is difficult to contact, because apart from laser welding, no other welding process can be used. Typically, mechanical processes, like insulation displacement contacts, crimping or bonding are used. The laser welding of copper wire often only works in combination with a skinning of the wire. Alternatives are insulation displacement contacts, a winding connection or crimping (Figure 3.86).

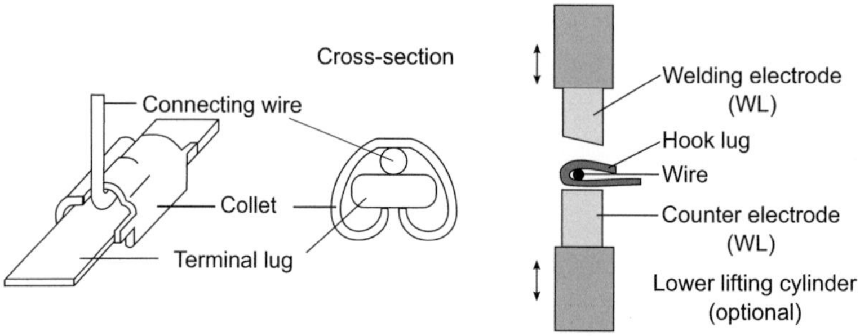

Welding

Figure 3.86 Clamping connection according to [DKI-00] and weld (Picture: Myac). Scan QR code to view demonstration video.

Aside from aluminium wire, resistance welding is universally applicable and therefore often used in the manufacturing of winding products. Crimping contacts will be examined more closely in the following, as these can be used for contacting pins. The wire is placed in a hook and clamped with the contact, using crimping pliers. This wire waste free method is very interesting in terms of its process dynamics and its economic efficiency. Wire park pins are not necessary here, as the contact is clamped directly to the bobbin. An approximately 8 to 12 mm length of free wire hangs from the wire guide nozzle and must be threaded or placed into the hook. An essential requirement for this application is the use of stable wires. Enamelled copper wires with gauges larger than 0.3 mm provide sufficient stability.

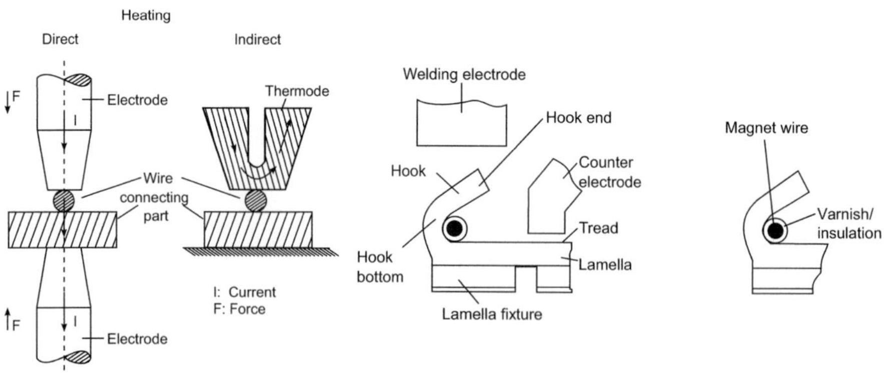

Resistance welding **Welding on hook collector**

Figure 3.87 Welding processes

Significantly thicker wires which provide a higher stability can be fixated to a contacting post or around it through a motion sequence. Hence wire park pins can be omitted as well. Alternative tools can be installed on the posts that cover these tasks. However, the terminated wires finally have to be loosened and removed and therefore this process is not waste free.

Contacting must ensure good mechanical and electrical contacting properties. Contacting processes with process times about 1.5 s are common in series production. Therefore, aside from considering the existing material combinations, the process sequence has to be designed accurately.

3.4.3 Bonding processes

Bonding describes the process of hardening the bonding enamel. The bonding enamel is applied to the copper wire as an overcoat. This is used for up to 80% of air coils to ensure the mechanical stability, which is necessary for the assembly of a coil carrier, the assembly of a coil core, the fixation for transport, or for the use of a mechanically stable air coil. Processes for bonding are differentiated by thermal and chemical reactions. Thermal processes can be realised by supplying hot air, storing parts in an oven or by electrical heating. However, using an oven is an exception due to its characteristics as a batch process with large process times. Processes with a chemical reaction are based on a wet-winding process. The wire is wound wet and a vaporizing solvent hardens the enamel. A quantitative comparison of the different processes has been performed in Table 3.6. A plus stands for a positive process property in terms of the criterion. Prerequisite for a homogenously bonded coil is a clean wire and preferably an orthocyclic winding, where the wires are parallel. For random windings, where the wires touch only pointwise due to the number of crossings, only minor stability can be achieved.

Tab. 3.6 Comparison of different bonding processes

Criterion	Hot air	Oven	Electrical Heating	Wet Winding
Process time	0	--	++	--
Bonding quality	0	++	++	+
Investment costs	+	-	0	++
Application range (dwire)	up to 0.4 mm	+	+	+
Degree of automation	+	--	++	++
Susceptibility to errors within the process	++	++	++	--

The hot air method is only applicable for thin wires and is generally hard to monitor, as the heat exchange takes place through the airflow. Due to turbulences, a precise temperature transfer to the wire is not always given. Accordingly, in-line process monitoring is not possible. For thin wires up to around 0.2 mm diameter, a relatively slow winding speed must be used in order to ensure a suitable heating of the wire. Typical speeds of winding spindles are around 500 rpm for average coil dimensions with 20 to 30 mm inner diameter.

The chemical process has the disadvantage that a solvent is needed which contamines the process environment and limits the winding process in terms of its dynamics. The use of electrical bonding has the advantage of short process times of around 2 to 5 s, as the copper wire, due to its electric resistance, warms up with a current density between 100 and 150 A/mm^2. Inline process monitoring and high repetitive accuracy as well as setting of defined temperature profiles is possible due to the consistency of the electrical resistance. Due to the heating from inside the wire, it can still be bonded after winding. When designing the process, one must keep in mind that heating the wire also leads to a higher resistance. In turn, during operation this leads to higher electrical losses, which leads to an increased warming of the part. This relation of temperature and resistance is shown in Figure 3.88. With growing quality requirements, this process is increasingly being used for thinner wires. To control the process, different control parameters can be used. If needed, voltage, current, or temperature can be controlled.

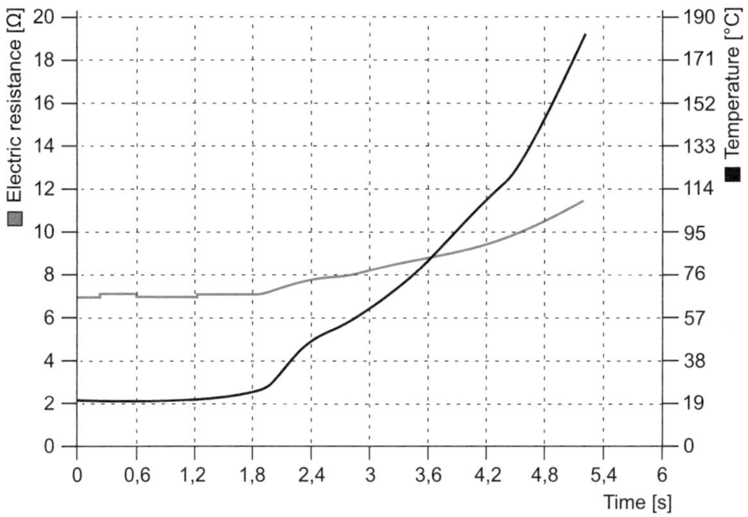

Bonding

Figure 3.88 Relation of wire temperature and resistance.
Scan QR code to view demonstration video.

The bonding process requires additional bonding enamel on the wires. Hardening can take place thermally or chemically. The large variety of coil types requires flexible winding processes, which can be adapted to the individual geometry circumstances. Ideally, air coils can be manufactured, for which the winding tool can be adapted to the coil geometry. The disadvantage of air coils lies in the downstream processes, where the winding has to be strengthened by bonding. Due to the missing termination, the possibilities of automated contacting are very limited.

3.4.4 Coil testing

The purpose of this section is to introduce testing procedures for series coil manufacturing.

Tab. 3.7 Overview of selected testing processes for winding goods

Product	Test procedure
Solenoid	– Impulse voltage testing – Partial discharge testing – Testing the insulation resistance – High voltage test: AC / DC – Resistance measurement – Inductance measurement
Armature/ Rotor	– Impulse voltage testing – Partial discharge testing – Testing the insulation resistance – High voltage test: AC / DC – Short circuited coil test / Search test – Resistance measurement – Inductance measurement
Magnet wire	– Impulse voltage testing – Partial discharge testing – Testing the insulation resistance – High voltage test: AC / DC
Stator	– Impulse voltage testing – Partial discharge testing – Testing the insulation resistance – High voltage test: AC / DC – Resistance measurement – Inductance measurement

Depending on the application, different processes for inspection exist. These are summarized in Table 3.7. The testing procedures for magnet wires are described in Section 2.3.1. The following section gives a brief introduction to the individual testing procedures. Testing is an important part of the process chain in order to sort out faulty coils.

For the impulse voltage or surge testing,impulse pulsed voltage is applied to the tested winding through a parallel circuit of a charged capacitor. This creates an electrical oscillating circuit, in which the power alternates between coil and capacitor. The oscillation is damped due to the electrical losses. This test indicates an amplitude curve characteristic for the winding inductivity, which always looks the same for faultless coils.

The partial discharge testing takes place with a defined testing sequence. A test voltage is applied to the winding. The voltage is increased until the first partial discharges take place. The according voltage is referred to as inception voltage. Subsequently, the voltage is decreased again until no further partial discharges occur. The according voltage is referred to as extinction voltage. Based on these two voltage values, a statement on reject is possible within a short testing time of a few milliseconds.

The insulation resistance check of a winding is performed by applying a voltage of up to 500 V DC. The current flow is measured through the insulation of the enamelled copper wire. By using a direct current, interfering electrodynamic effects can be neglected. Due to the winding capacity and the limitation to a small testing current of 3 mA, for example, this electrical test takes a bit longer, as the winding capacity has to be charged first. The insulation resistance provides information of the DC insulation strength.

The high-voltage testing with alternating and direct current serves to check the electrical insulation capabilities and voltage strength of air and creepage path on electrical parts or sub-assemblies. The size of the testing voltage, the duration and type of test as well as the maximum allowed leakage current are defined by national and international regulations and standards. In principle, the high-voltage testing is performed with the same procedure as for the insulation resistance testing.

Testing

Figure 3.89 High voltage testing (Picture: Eviro GmbH) and optical inspection. Scan QR code to view demonstration video.

The arc detection, also referred to as partial discharge measurement, describes the detection of micro discharges in a partial section of the insulation distance. As a part of the insulation system, these grow into increasingly larger defect areas. Depending on the stress and the type of defect, a full breakthrough may occur after hours, days, weeks, months or even years. The reason for this is the ongoing weakening of the insulation due to the constant micro discharges. The test is carried out using intelligent high-voltage testing of the insulation with different alternating current variants. Both inductivity and the electrical resistance are measured through a full bridge circuit with highly accurate components. The structure of this test circuit can be looked up in basic electrical measurement technology literature under the term Wheatstone bridge. Coil testing is specified by the product requirements. Depending on the application area, e.g. medical technology, there are detailed regulations for the documentation of component properties. Usually, coils are tested in terms of their resistance and their insulation capabilities.

3.4.5 Subsequent assembly processes

Aside from the necessary steps of pre-assembling the bobbin, several assembly steps are required after winding due to the use of coil in electromagnetic sub-assemblies, such as electric motors. The bobbin is usually supplied as an individual part or in bulk. Upstream processes, such as joining of end caps onto the lamination stack, are necessary in order to ensure the insulation between winding and lamination stack for the manufacturing of motors. Inserting contacts or contact pins as well as paper insulation on the sides of the segment is also possible. After winding, different configurations of the segments are possible. An example is shown in Figure 3.90. Pole chains, for instance, are intended to be assembled into a full lamination stack. Depending on the usage, single teeth can be assembled either on the inside or the outside of a stator, or as a pole pot. Depending on the product type, assembly parts are provided with the bobbin. In heating pumps, for example, T-segments are pressed into an end ring through a dovetail joint. Further examples of downstream processes are inserting additional insulation displacement contacts, or the assembly into housing, for example for valve coils.

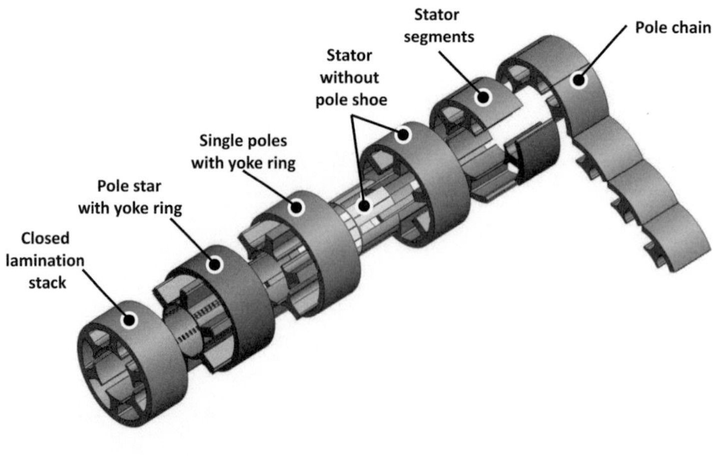

Figure 3.90 Typical mounting shapes of single elements for an EC-motor (Picture: Technoexpert Dresden). Scan QR code to view demonstration video.

The process chains for the assembly of individual segments are as numerous as the actual product requirement. The primary aim for a successful assembly is the shape accuracy of the coil after the winding process. Associated defects in the coil shape are discussed in the previous Section 3.3.11.

3.4.6 Secondary insulation

Secondary insulation describes measures of applying further substances to the winding, with the purpose of improving thermal, mechanical and electrical properties. For stators, this includes bandaging the winding head with an insulation tape, and the application of insulation resin into the spaces between the winding and the groove wall. Different processes, such as socketing, impregnation, trickling or over moulding, are available for the application of liquid insulation. These can be accelerated by means of positive boundary conditions, such as heat application with an oven, by applying power to the winding, or with UV radiation. Alternatively, the application can also be supported by a vacuum. Solid insulation can be realised as the assembly of phase separators in distributed windings, taping or bandaging of the winding head or the assembly of further insulation objects. A detailed introduction of the different material properties of existing insulation systems is covered in Chapter 2. Here, the emphasis is on the process properties for the manufacturing of winding products. For the selected materials, the product properties, such as high-voltage strength, partial discharge strength as well as the mechanical strength and thermal properties will be discussed.

The processing time is crucial for the choice of the manufacturing process for secondary insulations. This depends on the chosen material properties, such as duration for curing or jellifying, as well as the chosen application process. The process times and temperatures are specific for each material, and are illustrated in Figure 3.91. The process times of several application processes including their different phases are summarised in Figure 3.92.

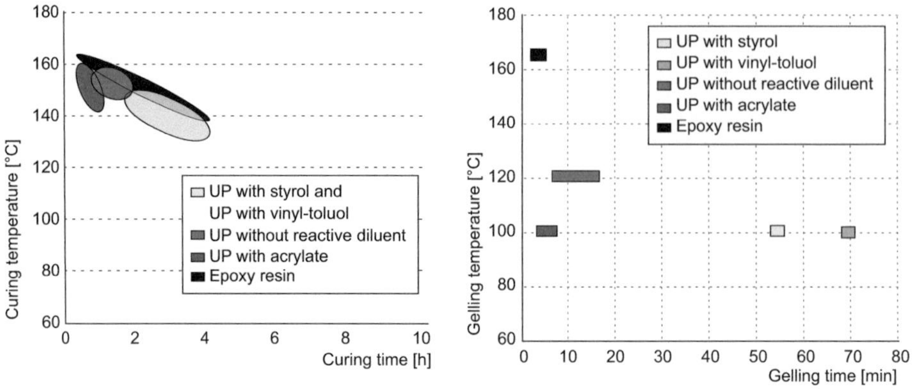

Figure 3.91 Typical curing times for secondary insulations

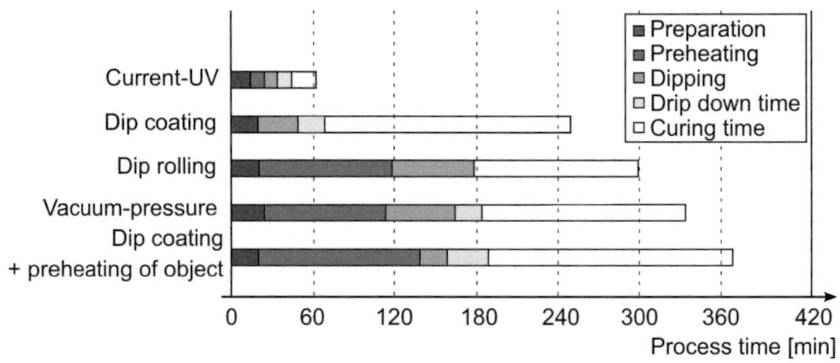

Trickle Insulation

Figure 3.92 Process times for different process applications.
Scan QR code to view demonstration video.

The application of liquid insulation resins is common in electric machine engineering. The impact of different resins on the partial discharge strength as well as the comparison of different process variants for the application of the resin demonstrate the degrees of freedom in process development, where process time, costs of manufacturing processes and the desired product properties must be considered. Examples are summarised in Figures 3.93 and 3.94. Deviations in partial discharge voltage between 3 and 5.5 kV are possible. In terms of the applicability of the secondary insulation, there is an interaction with the winding process. Flow paths must be provided for the resin application, which can be achieved by inserting precise canals. Accordingly, the quality of the secondary insulation also depends on the wire gauge and the winding scheme.

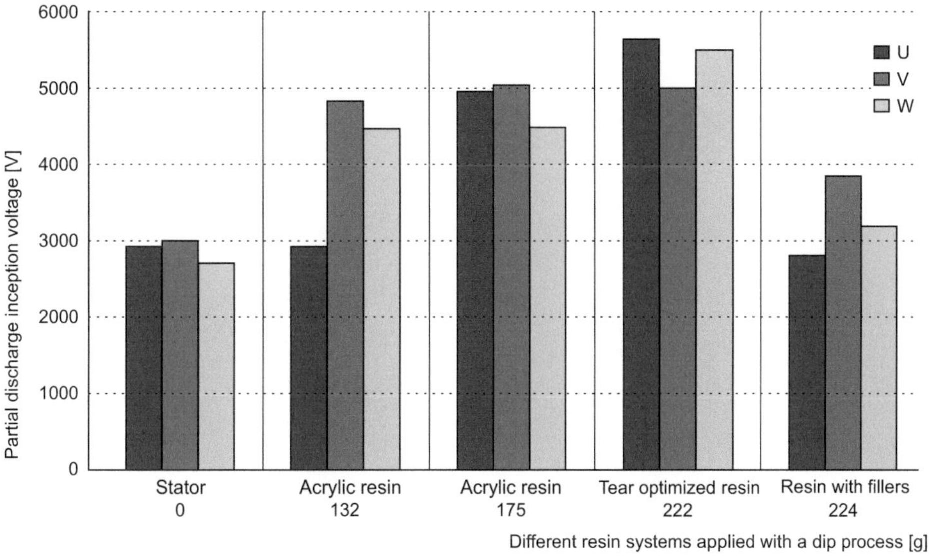

Figure 3.93 Influence of the resin system on the partial discharge characteristics [Ku-15]

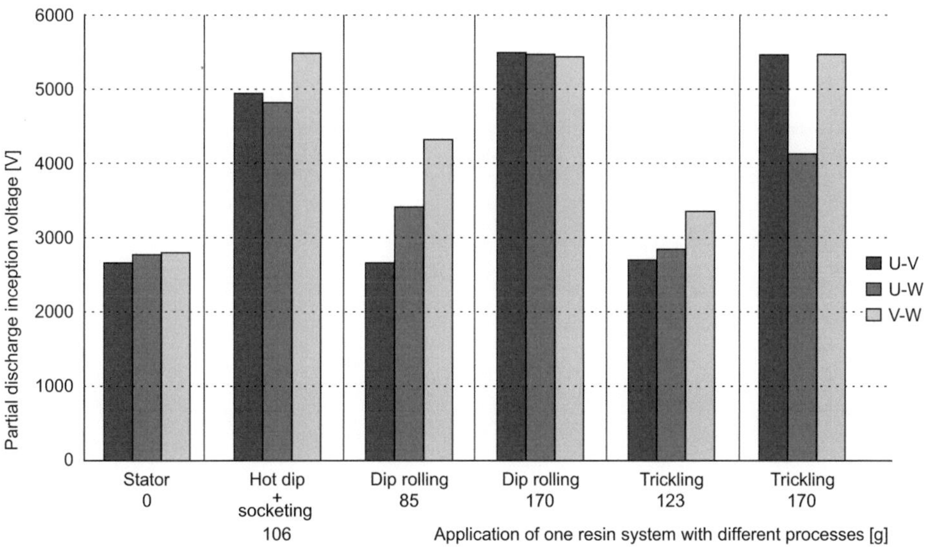

Figure 3.94 Influence of the application process on partial discharge characteristics [Ku-15]

The aim of the application is to fill possible cavities in the groove. Problems may occur if the adhesion of the secondary insulation to the wire can no longer be ensured due to a certain lubricant, or an increased lubricant application. Therefore, wires optimized for lubricants usually lead to weaker insulation characteristics. When heating the winding, an expansion could damage the wire insulation, and hence must be considered in the process design.

Figure 3.95 Secondary insulation based on plastic insert moulding

The use of secondary insulation increases the winding quality. It improves the winding's resistance to higher voltages or possible partial discharges and creates a mechanical protection. The choice of the insulation system is determined by the desired product properties. The process of manufacturing this system is characterised by the required process times and its costs.

References

[Dob-11] Dobroschke A: Flexible Automatisierungslösungen für die Fertigung wickeltechnischer Produkte. PhD thesis, FAU Erlangen. ISBN 978-3-87525-317-7

[Fe-13] Feldmann K, Spur G, Schöppner V: Handbuch Fügen, Handhaben, Montieren. 2nd ed., Hanser Verlag. ISBN 978-3446428270

[Ku-15] Kuschnerus M et al.: Elektroisolationssysteme für moderne Antriebskonzepte. Talk, Elantas Beck, 06.2015

[Küh-15] Kühl A: Flexible Automatisierung der Statorenmontage mit Hilfe einer universellen ambidexteren Kinematik. PhD thesis, FAU Erlangen. ISBN 978-3-87525-367-2

[Sel-15] Sell-Le Blanc F et al: Analysis of Wire Tension Control Prinicples for highly dynamic Applications in Coil Winding. Proceedings of the 5th International Electric Drive Production Conference, Erlangen. ISBN 978-1-4673-7511-5

[Tzs-90] Tzscheutschler R, Olbrisch H, Jordan W: Technologie des Elektromaschinenbaus. Verlag Technik, ISBN-10 3341008519

Automation

4

The aim of this chapter is to provide the reader with basic knowledge about the automation of processes, as well as the machines and facilities involved. It will also introduce the reader to different ways of linking processes and choosing the appropriate degree of automation. Lastly, it provides insight into aspects of economic operation for automated winding machines.

Figure 4.1 Automation technology in manufacturing

4.1 Introduction

The introduction provides a basic understanding of characteristic parameters, boundary conditions and the general procedure of automation. The cost considerations for different levels of automation are at the core of this section. The targeted audience has a background of project and manufacturing planning.

4.1.1 Definition of automation

In automated manufacturing, all operations, such as processing, control, handling of tools and workpieces, or quality monitoring, are performed mechanically, or rather electronically. The origin of the term is derived from the Greek word αὐτο and means independent or autonomous. According to DIN 72, "automation is defined as the equipment of a device, so that it works fully or partly without contribution by humans". For a better comprehension of automation technology, it is structured into degrees of automation.

Manual processes take place without the help of tools or machines. A mechanised manual process describes the operation with auxiliary tools. The mechanical processing chain can be divided into semi- or fully-automated processes. For semi-automated processes, there are often manual secondary operations, such as setting up or loading and unloading of bobbins, with winding as an automated main operation. However, the fully automated process includes fully automated secondary operations. According to this thought, the following degrees of automation for production technology can be derived: mechanical manufacturing, semi-automated manufacturing and fully automated manufacturing.

The key characteristic of automation is that beyond mechanisation, a control and regulation of the process can take place through a technical system. The advantage as opposed to mechanisation lies in the flexibility, which is given by the possibility of computer-aided manufacturing. However, the control is linked to high investment costs. Accordingly, automation is economically reasonable if the following boundary conditions are fulfilled:

- high production volumes
- product design suitable for automation with limited tolerances (bobbins, pins for termination, markings for data acquisition (barcode))
- parts suitable for automation
- high and constant product quality requirements
- manufacturing quality suitable for automation (bobbin).

Figure 4.2 shows a classification of exemplary functions for a winding product manufacturing line with the associated degree of automation.

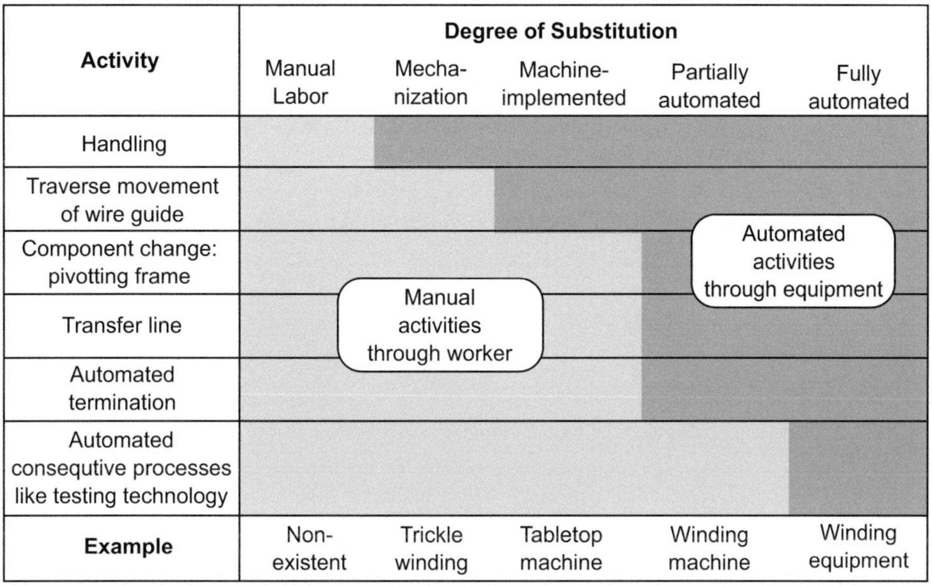

Figure 4.2 Degrees of automation in winding technology

The degree of automation can be chosen according to the application scenario. In mechanical manufacturing, the task of mechanised systems is to support human operations with technical auxiliary means, without technically implemented control or regulation of the process. This is suitable for small to medium production volumes and enables stable quality through its process reliability. Mechanisation is especially suitable for simple motion sequences for assembly processes. Production machines in semi-automated manufacturing have an electronic control. The control or correction of individual processes is implemented according to the closed loop principle. The transportation operation between machines, as well as loading and unloading, are at least partly performed manually. In fully automated manufacturing, the transportation operations as well as processes of loading and unloading are also implemented by automation. The controls of the machine take over the control or correction for all processes according to the closed loop principle. A control of every single process is possible due to the consistent automation chain. The human operations are limited to:

- maintenance
- raw material supply
- transportation of finished products
- monitoring the facility.

Of course, there are processes which cannot be automated economically due to their requirements. Instead, a manual or rather mechanised procedure is used, which can be performed on an integrated manual workplace. The decision for a certain degree of automation of each process step is based on an amortization calculation (Section 4.1.5). The assembly of components and sub-assemblies, which are difficult to handle, is an example for a process unsuitable for automation.. This includes loose wire ends of coils or limping components, as they often occur in cable assembly. Visual testing or joining processes are also limited to manual processes if component detection is difficult, e.g. due to a lack of orientation tags.

In order to achieve the described automation attributes, special functions have to be carried out. These include controlling, monitoring, notifying, displaying and operating. An exemplary process from winding technology is the control of the base wire tension during the winding process. The term *open loop control* includes the setting of machine variables with the aim of achieving a defined state, or rather a sequence of states of a technical process. As opposed to *closed loop control*, however, the properties of the sequence of states, or the process are not measured. An example is the winding routine with its variables feed rates, winding windows, number of layers and turns per layer. However, these variables are not measured during the process. The identification of deviations of physical values from the value of reference, and the automatic notification of the machine operator are called *monitoring* and *notification*. An example is the control of the wire tear during the winding process, or the test of winding readiness before the winding. *Displaying* and *operating* describes the visualisation of the process properties for the machine operator, which is often associated with the possibility of making changes via the operator panel. For example, typical testing characteristics can be visualised with a traffic light logic. Exemplary processes could be the joining of pins or the visualisation of camera recordings for the winding.

The automation system and its functions must comply with the specified requirements even under varying environmental conditions. An example from winding technology is the attainability of the coil target resistance with different deviations in the nominal wire gauge. As described in Section 2.1, these are standardised. The automated winding process must be robust, in order to balance out the deviation through a regulation or previously set counter measures, such as a structured bobbin surface. For varying parameters within the process window, the system must be able to react in time, so that the functionality is preserved. Larger deviations outside the process window are referred to as process malfunctions. These can only partly be balanced out by the automation system. To rectify the malfunction, the machine operator must be informed; he can then initiate counteractions. For an economic operation of the machine, a high system resistance against malfunctions and environmental influences is the primary target during the process design. Besides process reliability, however, safety precautions for humans, the environment, but also the machine itself must be ensured. This is achieved by measures including a spatial separation from the operator via a rotary arm, sliding doors or light grids.

Automation includes, aside from technical process properties and the technical product requirements, economic and safety-related aspects, which have to be considered in the automation concept. The process of automation also provides an opportunity to relieve human beings by taking over difficult or monotone tasks.

4.1.2 Criteria for the degree of automation

The choice of a suitable degree of automation for a machine or facility mainly depends on the boundary conditions of the product and the associated industry. The major users of winding products are found in telecommunications, industrial electronics, entertainment electronics, medical technology as well as the automotive industry (Sections 1.2.3 and 1.3.3). While the demand for winding products generally shows a slight upward trend, the increasing electrification of auxiliary drives in cars, as well as the automotive powertrain, is responsible for a significantly higher demand for winding products in the automotive industry. A boundary condition typical to the automotive industry is the high cost pressure, as the costs scale with the produced volumes. The product development cycles have become significantly shorter in recent years, also in comparison to other industries. Due to the required efficiency increase of each product generation, the associated innovation cycles have to become shorter. The product development initially focusses on the optimisation of products for their application. This leads to a rising product spectrum with smaller lot sizes for each product. As an approach to solve the challenge of increasing product variety with smaller production volumes, agile manufacturing was introduced, which can adjust flexibly to a varying production volume. Besides specialising products, the focus is also on increasing the performance, for example through higher fill factors, or miniaturising coils. In order to produce high tech products at different production volumes, the production is automated, while being kept flexible in terms of the product and the production volume through an intelligent layout of the machine (Section 4.2.2). The increasing integration of manual workplaces additionally increases the manufacturing flexibility (Section 4.3.2).

The challenge for European manufacturers lies in an efficient automation, in order to provide an alternative to manual manufacturing in low-wage countries. Further development of production technology is also necessary in order to gain competitive advantage compared to production technology from Asia, for example. Figure 4.3 summarises arguments for increasing automation in manufacturing.

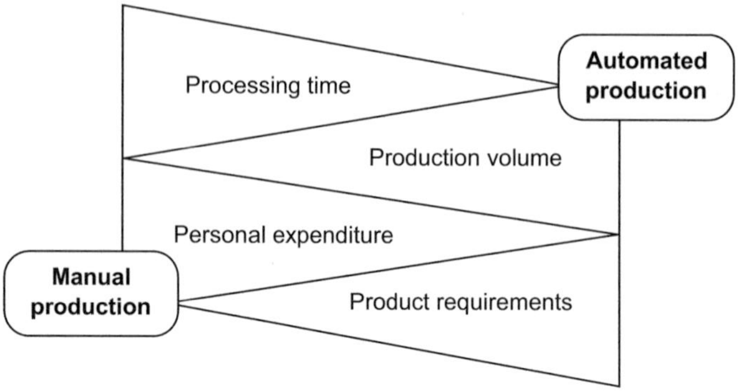

Figure 4.3 Reasons for automation

The aim of each manufacturing facility is the cost-effective operation of a production machine, or the operation of a product. The key figure for this is often the aggregated productivity, or profitability, which is defined as follows:

$$Profitability = \frac{Value\ added}{\sum Efforts} \tag{4.1}$$

Expenses include all materials and measures needed to manufacture a product. The *value added* describes transforming an existing item into an item with a higher value. Due to the high initial efforts for automation, this only pays off for high production volumes. However, it then allows for savings in labor costs, diminishing material consumption through efficient processes, or an increase in energy efficiency. This enables an economic and ecologic operation of the machine. The design of a manufacturing line takes place in different degrees of automation and can be set up in modules. These degrees of automation are introduced for winding technology in Sections 4.2.2 and 4.2.4.

Besides increasing efficiency, automation can provide better general and ergonomic working conditions. Simple, recurring tasks are taken over by machines. Operations involving health risks, such as lifting heavy parts as well as working in dangerous or unhealthy conditions (e.g. wire enamelling machines) are omitted. For certain tasks, automation is compulsory, for example due to requirements regarding technical cleanliness (clean room), high precision handling, such as in the assembly of micro parts, or due to hazardous components. Risks may occur for the worker due to high temperatures, for example after heating up self-bonding enamel coils, or the insertion of toxic materials. Some desired product properties, such as a certain surface condition, can also rule out manual handling. Controlling complex, technological processes, e.g. the wire guidance during coil winding with winding speeds of 30,000 rpm, is another exclusion criterion for manual operations.

Generally, this applies to a constant operation of machines with a high mechanical utilisation, which has high requirements on time production or maintaining a constant quality, such as achieving the highest possible fill factor. A reliable operation of machines is directly linked to the prevention of production downtime caused by human errors. This is ensured by continuous monitoring and control of process parameters, and results in high machine availability.

Because of the aforementioned shorter product cycles, production technology must enable a quick and cheap adaptation to changing market requirements. This is easier to achieve with a manual process. However, if the changes are linked to geometrical characteristics, these can easily be adjusted in fully automated controls via programming. If further changes are possible, depends on the chosen approach for machine design. The product design should already consider customers' demands, e.g. a flexibility for different types or product variants. This leads to higher basic investment for the production system.

High quality requirements for production or the product, as well as increasing performance in the automotive industry, lead to an increasing automation of winding technology in high-wage countries. An economic manufacturing can only be achieved with high machine availability.

4.1.3 Parameters for process automation

Different parameters were defined to characterise automated processes and their mechanical implementation, which will be introduced in this section.

A machine or process cycle time plays a central role in the design of interlinked production machines. It describes the quantitative target output of a system per time interval. As a planning value for the design, the theoretical cycle time for the rough planning of the line (Formula 4.3) and the cycle time (Formula 4.2) are distinguished to estimate the achievable production volumes per station. These are defined as follows:

$$Cycle\ Time = \frac{Working\ Time}{Output} \tag{4.2}$$

$$Cycle\ Time_{\,Theoretical} = \frac{(Working\ time - Setup\ time) \cdot Occupancy\ rate}{Output \cdot Utilisation\ factor} \tag{4.3}$$

For the theoretical cycle time, the operation time and set-up time have the unit minutes per shift. The production volume is given in pieces per shift. By taking into account the occupancy of the machine and the personal allowance factor, meaning the allocation of workers on additional machines, the theoretical cycle time gives a more precise planning value.

A central unit for the reliability of a machine is the technical availability, which is also referred to as utilisation factor. "The availability of an object describes the probability of this object to be found in OK condition at a certain point in time" [Lan-10]. This is described in VDI standard 3423. "An availability of 99.8 % says that a machine statistically works for 99.8 % of its operating time" [Lan-10]. To determine the technical availability, the mean time between failure (MTBF) and the mean time to repair (MTTR) are used. The mean operating time describes the average period between to failures of the machine. An alternative figure is the mean failure rate, which is the reciprocal of the operating time. The repair time describes the average time required to fix a problem. The technical availability V_T results from the two times described in Formula 4.4.

$$V_T = 100\% \cdot \frac{MTBF}{MTBF + MTTR} \tag{4.4}$$

The overall availability of a whole machine V_{Ges} results from the availability of individual components V_n. It has to be considered, whether the number k of the components is arranged in series or parallel. The coherence for a series arrangement $V_{Ges,Ser}$ and a parallel arrangement $V_{Ges,Par}$ can be described according to Formulas 4.5 and 4.6 respectively.

$$V_{Ges,Ser} = V_1 \cdot V_2 \cdot V_3 \cdot \ldots = \sum_{n=1}^{k} V_n \tag{4.5}$$

$$V_{Ges,Par} = \frac{1}{V_1} \cdot \frac{1}{V_2} \cdot \frac{1}{V_3} \cdot \ldots = \sum_{n=1}^{k} \frac{1}{V_n} \tag{4.6}$$

A characterisation of the overall machine availability is possible using the individual machine component availabilities and following the same schematic. These formulas for series or parallel order are already known from electrical engineering. Accordingly, the availability of a series arrangement is determined by its weakest link. Therefore an increase in the number of components per machine is linked to a decrease in availability. This is why the availability of a multi-spindle machine is always lower than a single-spindle machine. In the event of malfunction of one spindle, the whole machine usually needs to be stopped and repaired. The alternative of switching off individual machine components is often impossible due to the functional dependencies within the machine. The availability can also be influenced by a more robust design of the machine components. That is why machines for heavy-duty wires typically have a better availability. Aside from a higher number of machine components, a higher degree of monitoring can lower the availability. On the one hand, this is due to additional testing components and, on the other hand, due to lower process reliability. Based on an empirical analysis of component availabilities for a linear winding machine, an exemplary availability of the machine modules was determined. A series arrangement was assumed. The results for the overall machine availability are shown in Table 4.1. It is evident, that a reduction of the secondary process and machine components always results in a better availability of the machine.

Table 4.1 Derivation of a machine availability on module level

Machine module	Module availability	Overall machine availability
Spindle	99.53 %	
Wire tension control system	99.36 %	
Traverse system	99.18 %	
Control	99.41 %	**97,12 %**
Safety equipment	99.98 %	
Wire cutting equipment	99.63 %	

A statistical quality description of the manufacturing process uses different parameters for processes and machines. These are used for a static process regulation, which allows for a parameter adjustment during the process, so that the defined product parameters can be achieved although the input varies. A distinction is made between random influences, e.g. temperature changes or the positioning precision of a machine, or influences of systematic origin, like a tool wear or ramp-up the processes. During production, random samples are taken and compared in terms of the allowed manufacturing deviations. The findings of this analysis are used to determine the machine capability c_{mk} and the process capability c_{pk}.

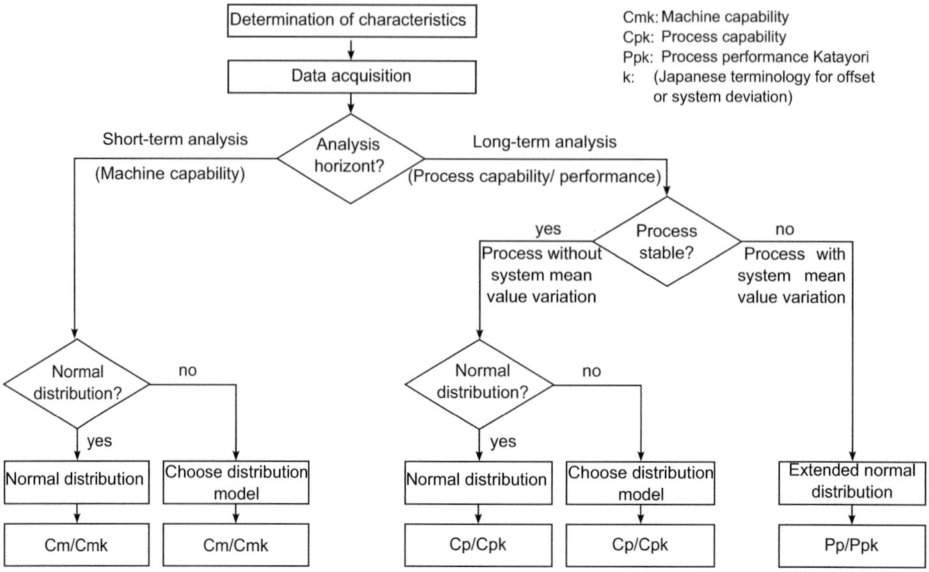

Figure 4.4 Procedure for determination of characteristic values [Bo-04]

The machine capability describes the quality capability of a machine under ideal conditions and is a measure for the short-term capability. The influences from materials, human behaviour, methods or environment are kept constant (Ishikawa diagram/winding errors, see also Section 3.3.11). A suitable observation period consists of approximately 10 random samples taken from 50 consecutively manufactured parts. The process capability describes whether the process can fulfil its requirements during production and is therefore more strictly regulated than the machine capability. As opposed to the machine capability, thc process capability is determined under real process conditions. The observation period for this is at least 20 days taking partial samples with sizes of 1 to 30 pieces. The procedure to determine these values is displayed in Figure 4.4. Due to the focus on coil winding technology, only c_{pk} and c_{mk} values with standard distribution are relevant, as these are valid for technically controlled or stable processes. A process with varying mean distribution, as is the case for the p_{pk} value, would not be chosen for an interlinked, highly automated production line, as this would lead to an unjustifiable amount of rejects. Typical characteristics to determine the machine and process capabilities are strongly product dependent. For the winding process, typical characteristics are the number of turns, the electrical resistance of the coil, the wire gauge or the coil dimension. For contacting processes, the parameters depend on the actual process. For resistance welding, for example, the sink-in depth, the welding time and the welding current can be monitored and evaluated. Typical monitoring units for assembly processes are force-stroke monitoring or the monitoring of torques. In testing technology, the electrical testing can be performed with the previously mentioned resistance. For optical tests, distances and positions are critical characteristics. For the tactile test, the repetitive accuracy is a capability characteristic. In order to calculate each capability, the following equations can be used. Accordingly, the capability results from the minimal difference between the upper specification limit *USL* and the mean value *X*, divided by three times the standard deviation σ and the difference of the lower specification limit *LSL* and the mean value, divided by three times the standard deviation σ (Formulas 4.7/4.8).

$$c_{mK} = \min \left| \frac{USL - \bar{X}}{3\sigma}, \frac{\bar{X} - LSL}{3\sigma} \right| \tag{4.7}$$

$$c_{pK} = \min \left| \frac{\bar{X} - LSL}{3\sigma}, \frac{USL - \bar{X}}{3\sigma} \right| \tag{4.8}$$

Typical values are 1.33, 1.67, 2.0 or 2.33, due to the normalisation on three times of the standard deviation. A comparison of c_{pk} values and reject, or failure rates in ppm (parts per million) is displayed in Table 4.2. Typical technical limits for the capability of winding processes c_{pk} are between 1.33 and 1.67.

Table 4.2 Relation of process capability and reject rate [Br-11]

Process capability	0.33	0.67	1.00	1.33	1.67	2.00
Standard deviation	1σ	2σ	3σ	4σ	5σ	6σ
Good parts in %	68.26	95.46	99.73	99.9937	99.9999943	99.99999998
Rejects in %	31.74	4.54	0.27	0.0063	0.000057	0.00000002

Especially in the automotive industry, increasing requirements demand for process capabilities with values of up to 2.0. This complicates both the process design and the technological/ mechanical implementation of the machine. However, for winding products it is still possible. Nevertheless, the higher the process capability, the higher the expenditures for automation and machine costs. In order to motivate a customer to increase c_{pk} to 2.0, they must be sensitised for the higher technical availability of the machine during development.

To evaluate the distribution function, which results from the random capability analysis sampling, the terms capable and controlled are used. If the distribution of different random samples shows a stable mean value, the process is called controlled. If a process is controlled, it process can statistically be compared to the given tolerances. If the desired characteristics are within the tolerances, the process capability is given. This is illustrated in Figure 4.5.

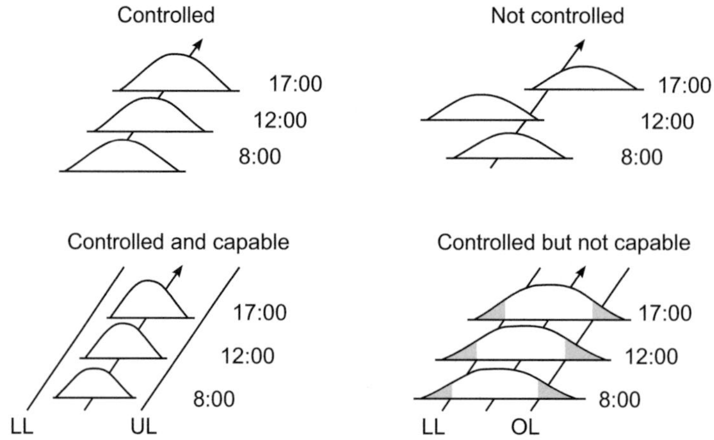

Figure 4.5 Process control and capability

Another parameter, the reject, can be derived from the capability. It describes the amount of produced parts, which, due to quality issues, cannot be processed further, or only after rework due (Table 4.2). Additionally, the internal reject rate is defined. It describes the produced parts, which the machine produced as rejects, despite the fact that the quality of the semi-finished products was adequate. The source of error therefore lies within the process or the machine. Typical internal reject rates are between 0.5 and 1 %. A central parameter to describe a production system is the overall equipment effectiveness (OEE).

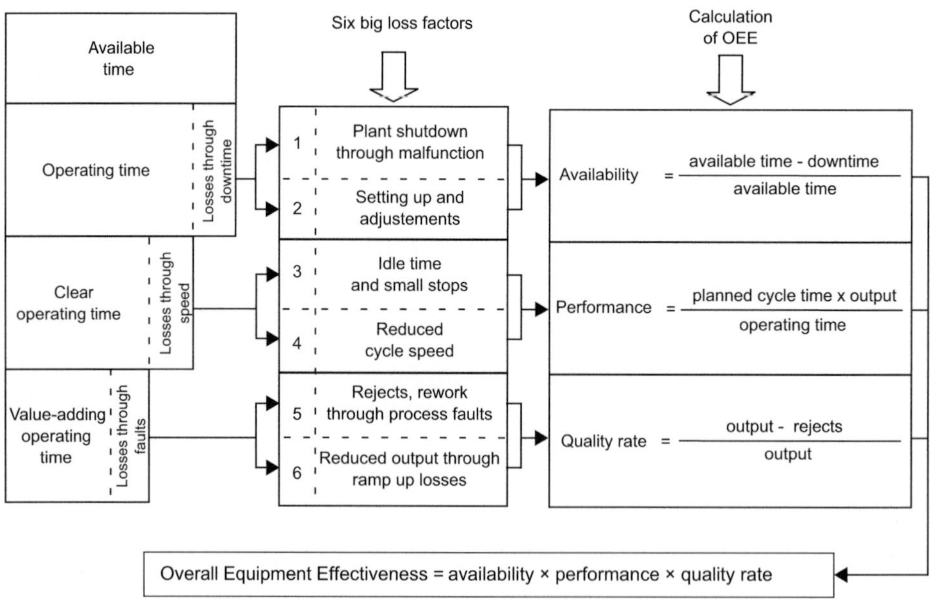

Figure 4.6 Calculation of the OEE

It is a measure of the value added of a production system in terms of productivity and losses. OEE takes three central factors into account: the capability, the performance rate and the quality rate. In turn, these depend on six central loss factors. A schematic to determine the OEE is illustrated in Figure 4.6. Accordingly, the basis of determining the OEE is measuring the most important times for each type of operation and the amount of reject with the associated output. An approach to optimize the overall equipment effectiveness is total productive management (TPM). Its purpose is to prevent losses and waste, in order to achieve a production with zero defects, zero breakdowns and no quality losses, while completely preventing accidents. A possible approach to increase the productivity, for instance, is increasing the value added by complementing further manufacturing steps, or the reduction of machine errors by installing more robust components. The quality goal can be achieved by reducing the process errors, for example through a better regulation or a reduction of quality issues. The costs can be achieved by reducing the maintenance costs, for example by installing energy efficient components, or by reducing the operating personnel with a higher degree of automation. Similar measures are possible for the areas of delivery, safety and employee morale and lead to an increase in OEE.

In the practical design of automation solutions, quality and cycle time are the most important parameters. For the evaluation of an automation solution, a clear definition of the considered quality characteristics for c_{pk} and c_{mb} values is necessary in an early stage. Especially when choosing the right manufacturing process for the desired task.

4.1.4 Different phases of automation

In order to enable an efficient automation, different steps of product manufacturing must be considered. In the first phase, the necessary manufacturing steps for the product are defined. Subsequently, a detailed plan of the automation takes place in the second phase. While doing so, each manufacturing can be divided into three steps itself: preparing operations, the manufacturing step and the subsequent testing or rework. For the winding process, these three steps consist of testing for winding readiness, winding and testing of the winding. The testing can be realised as an optical test with the shadowing principle. For the magnet assembly, the chain can be defined with these steps: applying the adhesive (preparing operation), magnet assembly and testing of the magnet position using a camera. For the individual main functions, different processes are available. In addition, even secondary processes, such as handling or part supply can take place in various ways. For winding products, the process chain described in Chapter 3 is applicable for automated manufacturing.

- Part supply: palletiser, vibration conveyor, step feeder, …
- Part handling: articulated-arm robotics, Scara robotics, portal robotics, …
- Assembly: joining, locking, spring assembly, …
- Winding linear winding, flyer winding, needle winding, …
- Contacting fork welding, sleeve welding, loop welding, …
- Joining: screwing, orbital riveting, pressing-in, …
- Testing: tightness testing, functions testing, DMC scan, …
- Labelling needle imprinting, inkjet printing, pad printing, …

Figure 4.7 Typical processes for winding products

For each process step there are therefore different processes and machines to fulfil the task. Several typical solutions for the process chain are shown in Figures 4.7 or 4.8 as pictograms above.

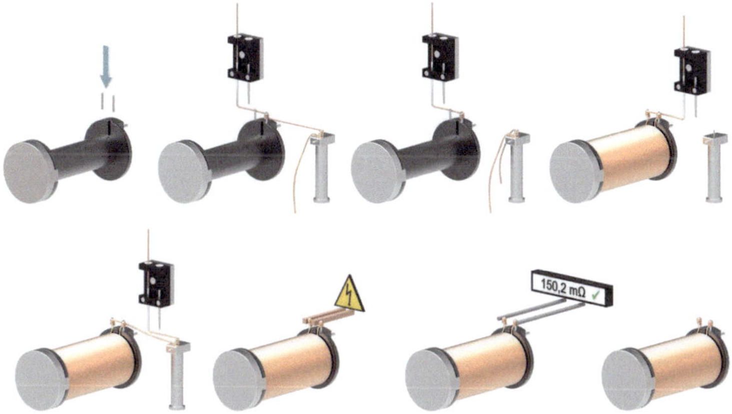

Winding products

Figure 4.8 Typical process steps for winding products. Scan QR code to view demonstration video.

Special machine engineering or rather winding technology tends towards the delivery of complete solutions for different products. Therefore, the upstream and downstream processes come from the same supplier. Product design in particular sets requirements, which require an integration of multiple tasks into a manufacturing chain or machine. The overall focus on the process chain results in more degrees of freedom for a manufacturing concept, compared to a design based on single manufacturing modules.

4.1.5 Amortisation scenarios

The aim of selecting the degree of automation is to choose a solution, which results in the lowest unit costs when considering all costs. In order to determine the unit costs, the following cost aspects have to be considered:

- acquisition costs incl. interests
- costs of energy and compressed air
- planned and unplanned maintenance costs
- installation space costs
- set-up costs
- personnel costs
- material costs for wire and bobbins
- reject costs

In order to enable a comprehension for the cost relations and possible unit costs, or cycle times for different degrees of automation, the following production scenarios were considered. These are set ups for typical applications depending on the winding process and for different degrees of automation (Figure 4.2) as summarised in Table 4.3. The degree of automation indicates the relation of manual and mechanically self-acting tasks and processes. It was assumed that the personnel costs per person increased with a higher degree of automation, because the qualification needed to work as a machine operator must be higher. Cylindrical coils used for the linear winding, as described in Chapter 1, are chosen as the reference. For a better comparison, flyer winding uses the same reference part. Needle winding, on the other hand, is typically used for closed lamination stack stators. Significantly higher process and cycle times result from the efforts for winding multiple coils and phases.

Table 4.3 Comparison for different degrees of automation
(Machine concepts with parallel times = 2-times pivoting frame/rotary tables)

Winding process/ object	Degree of automation	Process	Process time in s	Cycle time in s	Equipment costs in k€
Linear winding/ solenoid	Mechanical	- Winding with 1 spindle	120	120	40
	Partially automated	- Winding with 4 spindles - Manual loading/ unloading	15 30	30	200
	Fully automated	- Winding with 4 spindles - Automated loading/ unloading	7.5 7.5	7.5	250
Needle winding/ full lamination stack	Partially automated	- Winding with 4 spindles - Manual loading/ unloading	60 60	60	350
	Fully automated	- Winding with 4 spindles - Automated loading/ unloading	47.5 10	47.5	450
Flyer winding/ solenoid	Mechanical	- Manual Termination - Winding with 1 spindle - Manual taping - Manual wire cutting - Manual loading/ unloading	10 10 10 5 7.5	32.5	100
	Partially automated	- Winding with 1 spindle Manual loading/ unloading	10 7.5	10	200
	Fully automated	- Winding with 2 spindles - Automated loading/ unloading	5 5	5	350

Due to the difficult accessibility in the winding machine, caused by the flyer arms, flyer winding machines are typically chained to other stations or processes using a rotary table (Section 4.2.2), while a linear winding machine can integrate the stated functions into its machine design. Based on these assumptions and knowing the values for the described costs above, an amortisation calculation can be established, as shown for flyer winding in Figure 4.9. Different procedures and approaches are possible. The total costs for each degree of automation are plotted in Figure 4.9 for a flyer winding production line and over the produced volume. For each solution there are different steps, which are necessary due to the addition of a new shift or a new production machine, in order to achieve the target volume. For the comparison shown, the critical production volumes of a fully automated line are not achieved, which is why it remains constant.

Generally, the amortisation calculation is an evaluation process of investments, which considers the period of full return on the invested capital, in short, return on investment (ROI). A typical amortisation period is between 1 and 5 years. In general, there is a distinction between static and dynamic calculations.

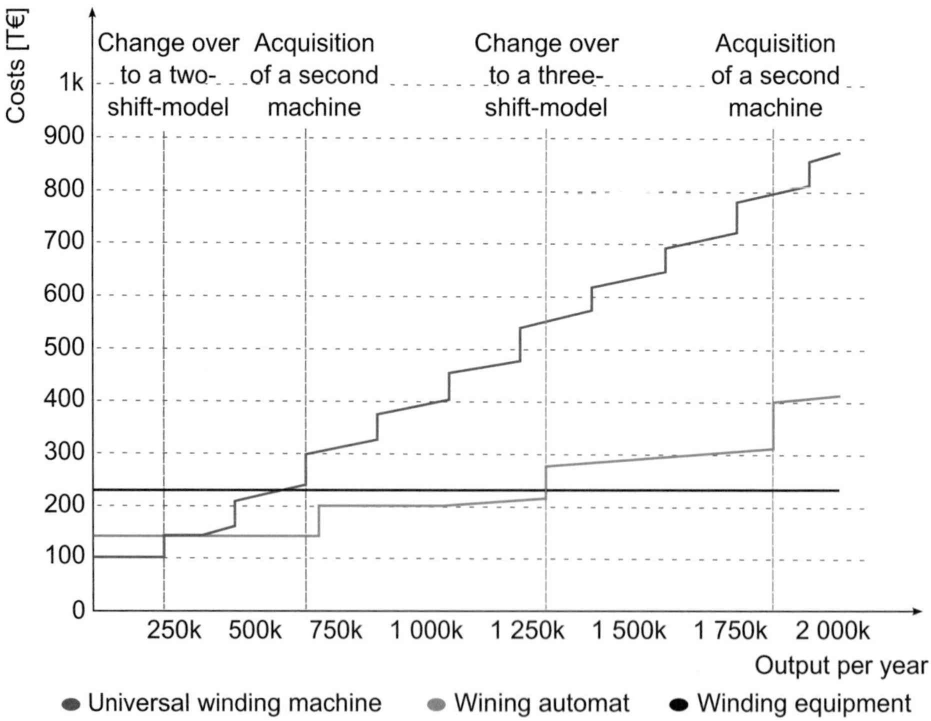

Figure 4.9 Amortisation calculation: Overall costs against output

The static calculation determines the point in time, at which the expected cash flow C_t equals the initial investment I_0

$$I_0 = \sum_{t=1}^{n} C_t \tag{4.9}$$

The dynamic calculation determines the point in time, at which the cash flow discounted with the relevant interest rate i equals the initial investment. This is the procedure used in economy.

$$I_0 = \sum_{t=1}^{n} C_t \cdot (1+i)^{-t} \tag{4.10}$$

Accordingly, the overall costs divided by the revenues per year gives the amortisation period. The machine-hour rate is often used to plan the running costs and to understand the cost distribution per machine. This is illustrated in Figure 4.10 for different degrees of automation of flyer winding technology.

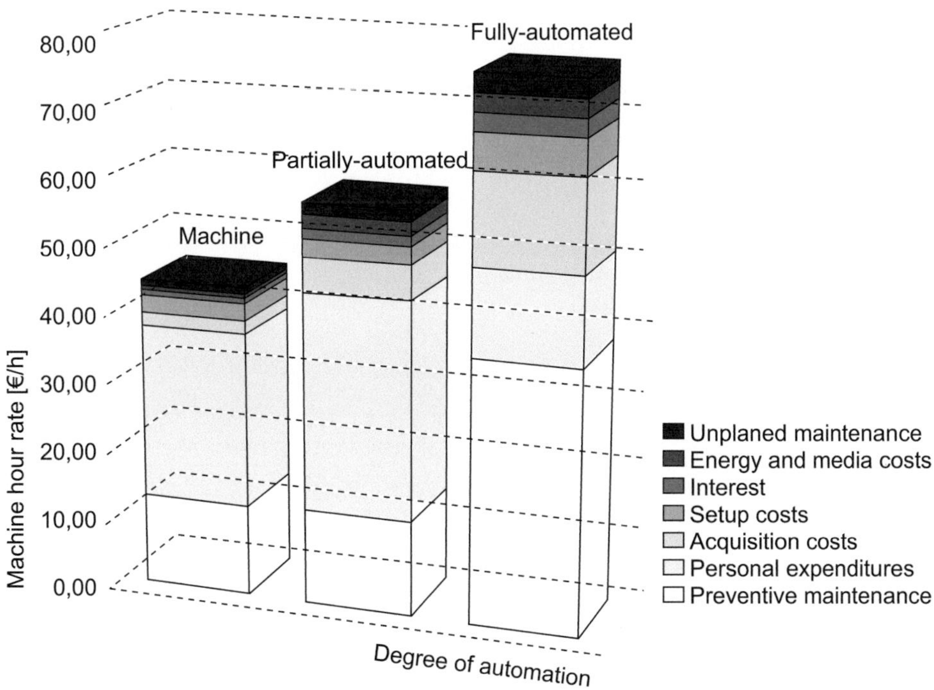

Figure 4.10 Machine hour rate for flyer winding machines

As expected, the machine investment and the follow-up costs increase with a higher degree of automation. The relation of investment costs and maintenance costs to the running costs for personnel almost inverts with higher degrees of automation. A fully automated machine has a cycle time of less than 6s. A semi-automated machine enables cycle times of 15 to 30 s. The mechanical process usually has a cycle time of longer than a minute.

Apart from the output of the individual station with its costs, the whole process chain is also a relevant factor in the amortisation calculation. In an automated manufacturing line, the winding process is usually the one with the longest cycle time. Using the example of a stator with a low fill factor and 12 poles, this can be shown very clearly. A needle winding machine can only wind this stator with one phase and needs an investment of around 350 k€ with a cycle time of around 60 s. A comparable open pole chain winding machine with multiple needles can wind the whole stator with all phases in 80s, however, its investment is around 525 k€. When comparing the cycle times for three phases, the open pole chain winding machine with 80 s is more than twice as fast than the single needle winding machine, which takes 180 s. Using a 4-spindle setup for the pole chain winding machine it could achieve a cycle time of 20 s, which increases its advantage even further. However, the downstream processes of the chain winding machine include confectioning of the wires (costs of around 40 k€), rounding and supplying contacts (60 k€), as well as contacting (80 k€) and inserting of an end ring into the housing (50 k€). The only comparable downstream process for the needle winding machine is the contacting of the wires (80 k€). Accordingly, the chain winding machine costs are significantly higher at around 750 k€ as opposed to the needle winding machine with 330 k€. However, the pole chain winding machine has a very good cycle time and enables higher fill factors. The advantage of choosing the needle winding machine with a closed lamination stack rather than the single-tooth design for the chain winding machine, is that considerably less efforts for contacting are needed, which means in lower costs. In this case, a simple trade-off between the winding stations would not have evaluated the follow-up costs due to the downstream processes correctly, and would have led to a false result.

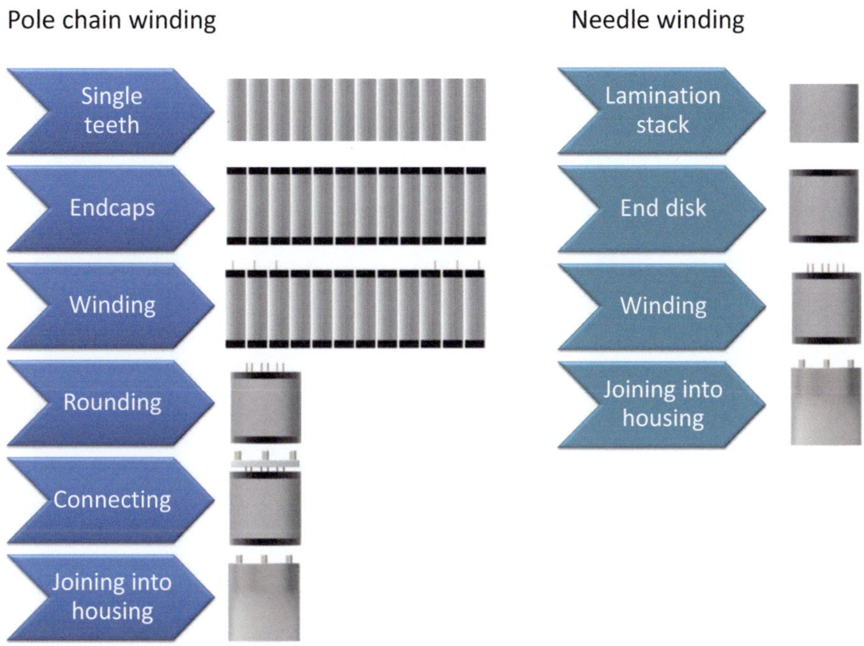

Figure 4.11 Process chain comparison for needle winding and pole chain winding

For the automation of handling tasks there is an empirical rule stating that the machine investment should not exceed the corresponding personnel costs for one year. Otherwise, the task is usually not automated.

The consideration of the cost-benefit analysis and amortisation scenarios is usually the responsibility of the operating company. Depending on the industrial background, different tools for the evaluation are used. These range from considering the total cost of ownership (TCO) in the automotive industry to a dynamic amortisation calculation for industrial suppliers.

4.2 Automation Concepts

Figure 4.12 Rotary table with robots and palletiser

The following section gives an introduction to the different technical aspects of automation technologies during machine design and production layout. Possible configuration levels of assembly machines, different feeding systems and interlinkage concepts as well as the available degrees of automation for winding machines and part properties relevant to automation are introduced. This section is therefore directed at process and layout planners for production machines.

4.2.1 Industrial assembly concepts

The smallest unit in an automation concept is the processing station. In winding technology, or in assembly technology in general, the assembly machines are classified according to their functions. An exemplary distinction between assembly stations, assembly cells, assembly centers and assembly systems is illustrated in Figure 4.13.

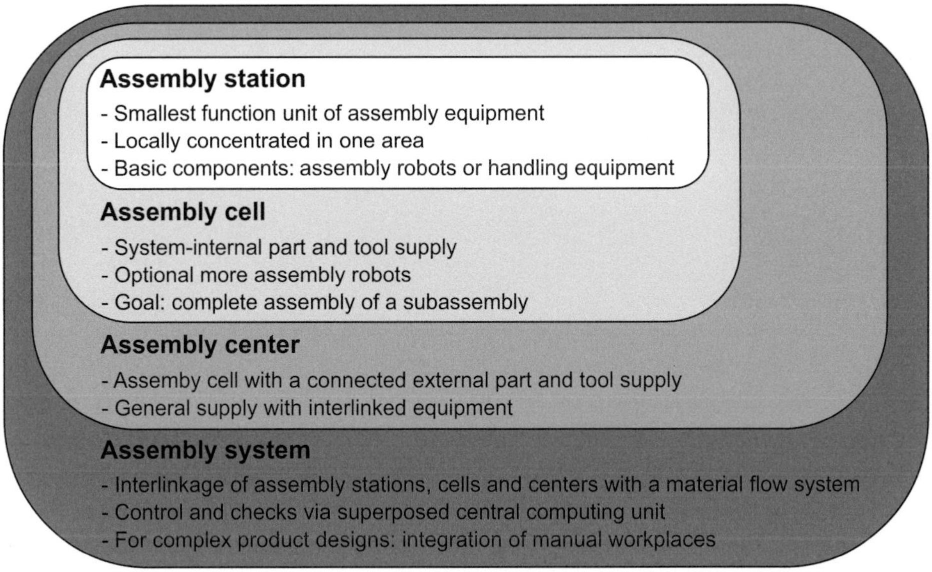

Figure 4.13 Upgrading stages for assembly machines according to [HES-93],[HES-00]

As the focus of this chapter is on automation, the assembly station is not discussed any further. Instead, the configuration levels after the assembly cell will be covered. If an assembly cell is linked to an external storage for assembly jobs, assembly fixtures and assembly tools, this unit is referred to as an assembly center. However, the storage is out of reach of the assembly robots. All of the material and information flow to the storage is controlled and monitored. An exemplary setup of assembly center components is illustrated in Figure 4.14.

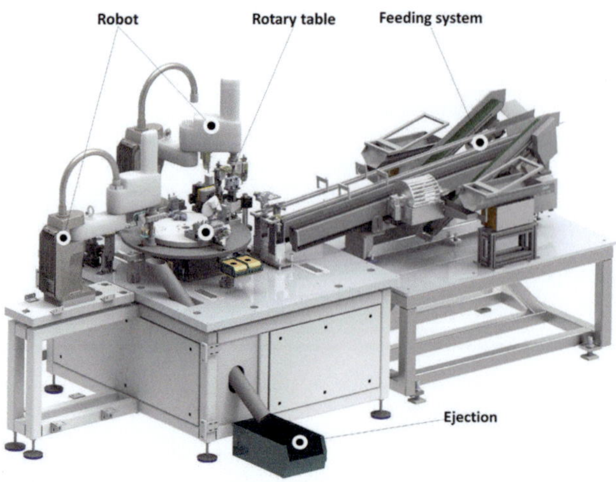

Assembly center

Figure 4.14 Components of an assembly center including.
Scan QR code to view demonstration video.

Assembly systems are interconnections of different assembly machines, such as assembly stations, assembly cells and assembly centers. Analogous to manufacturing systems, the material flow is also automated in assembly systems. The use of assembly robots is common, although manual workplaces can be included, as complex assembly processes cannot be automated, or only at very high costs. An example of electric motor assembly is illustrated in Figure 4.15.

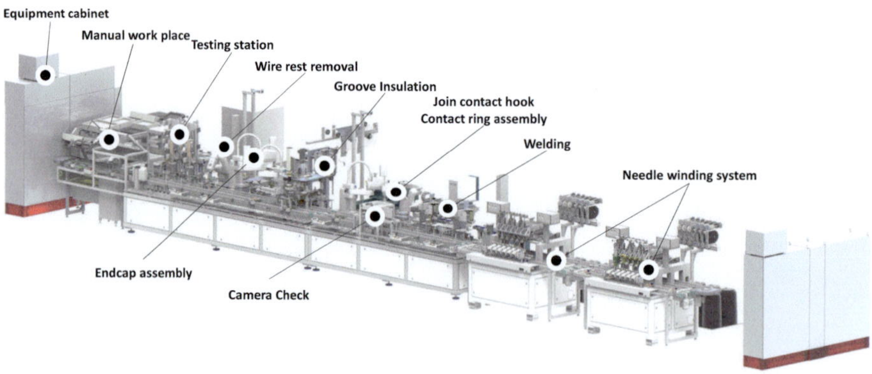

Figure 4.15 Example of an assembly center for electric motors

Analogous to manufacturing system, a high degree of automation in an assembly system usually leads to lower flexibility. This means that a machine can assemble a certain product very well and quickly, but is also limited to the specific product features. In production systems, there are conceptual differences between flexibility and adaptability. Flexibility is the reaction to a pre-set or planned event, for example a variation in stator length within the same design. Adaptability means adjusting to an unforeseen event, such as a completely new stator design. Flexibility is achieved by a modular system of fundamental elements and basic units for both the mechanical system as well as, and even more important, for logistics and control technology. This is the only way to configure individual solutions from a standardised modular system. Extruded and anodised aluminium profiles have been established as fundamental elements for frames, fixtures, transportation sections and superstructures. They provide the necessary mechanical stability, are corrosion-resistant and easy to process. Often they are available in multiple sizes and profile cross sections and therefore enable an optimal adaption to the requirements with minimal use of material. With the appropriate removable connections, transfer systems can be individually adjusted, changed or expanded on demand. Later re-use of the building elements due to changes or modifications of the system plays a vital role in terms of costs. Figure 4.16 shows that flexible assembly systems mainly compete with traditional manual and hybrid systems. Therefore, they are best suited for smaller and medium production volumes. For very high production volumes, automated and NC-controlled or cam-controlled assembly machines are still used. This is because the advantage of the higher speeds due to higher throughput weighs heavier than the disadvantage of the lacking flexibility.

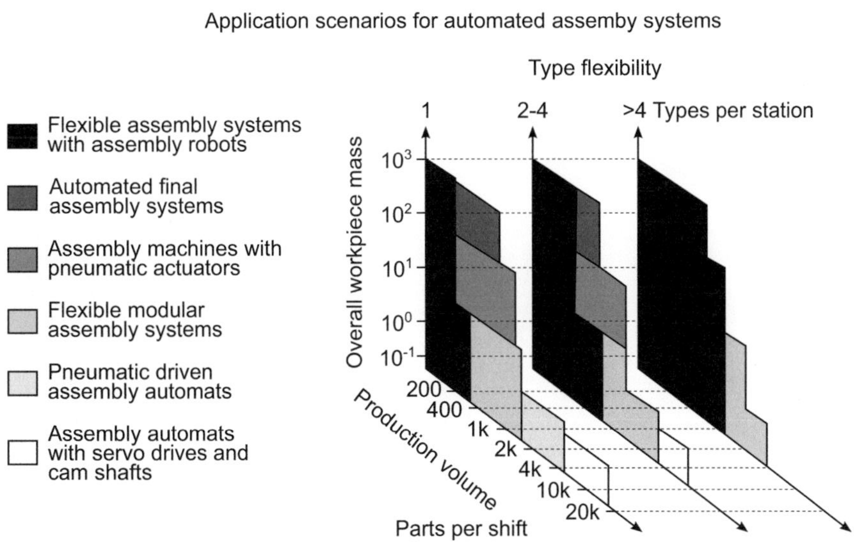

Figure 4.16 Application scenarios for assembly systems according to [HES-93],[HES-00]

4.2.2 Conveying systems and layouts for manufacturing lines

During conceptual design of manufacturing lines, the primary efforts focus on arrangement of material flow and assembly stations for an optimal supply of production material to and from the machine, and the best accessibility for the operator or the most space saving layout of a fully automated system. Nowadays, all concepts share the modular structure of a manufacturing line. Accordingly, machines can be moved, replaced or multiplied, if the production scenario requires it. The basic motivation for introducing modular structures is the demand for an increasing number of variants while the production volumes of individual variants decrease. An area of conflict results between the implementing efforts, the standardisation of individual modules and a limited flexibility. In general, the system is divided into smaller units by modularisation. These modules are functionally and physically independent units and fulfil at least one function. If the range of system functions increases, the number of modules grows. Since the complexity and demands of the system require higher lead time, it takes longer before one can profit from the advantages of modularisation. However, advantages of modularization are the parallelisation of development tasks, a simple derivation and configuration of system variants, and the potential to reduce the complexity of a system and the efforts for change of the system. This results in a higher number of variants, while simultaneously decreasing the development time of individual variants. The disadvantage of this approach is the similarity of products and the fact that the modules may not comply with all customer requirements. [Rot-14]

A typical setup for an automation concept is illustrated in Figure 4.17. The layout shows a line layout, where the individual stations are connected in series, with manual work places in the free spaces in-between. These provide the benefit of possible stations for rework, which is necessary during the setting-up or in case of failures of the process. In addition, if a station malfunctions, the function can be taken over manually, so that the whole production is still running with reduced speed.

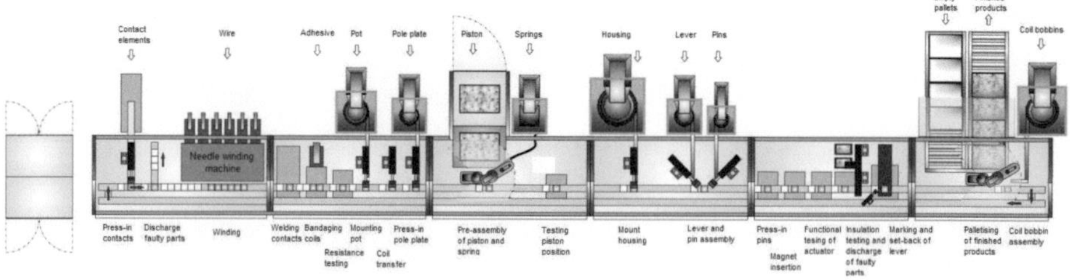

Figure 4.17 Typical line layout

Generally, three possibilities for the setup or the chaining of a manufacturing line layout are available; these are illustrated in Figures 4.18 and 4.19 respectively.

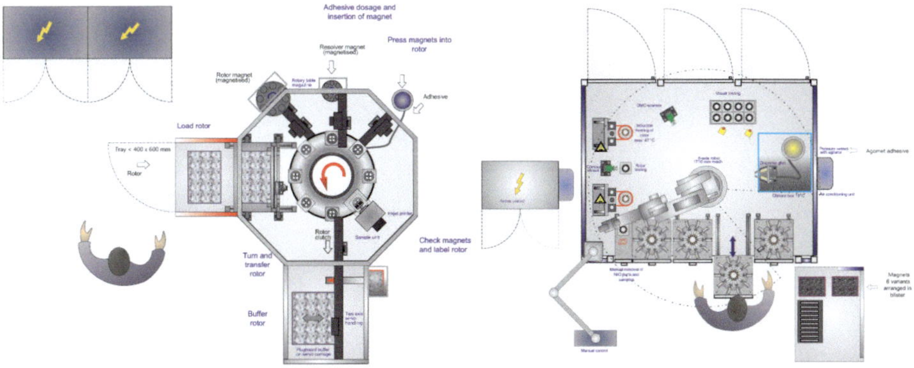

Rotary table **Manufacturing / robot cell**

Figure 4.18 Layout concepts for feeding systems – rotary systems

The first system, a rotary table, is very quick due to its rotational motion, but provides only a limited number of stations. The accessibility for machine maintenance and to the part is limited as well, if it is not supposed to be handled separately between each station. This linking technology is extremely suitable for the realisation of parallel processes. The second system, the manufacturing cell, is a structure that comprises multiple processing stations, which are loaded and unloaded with a robot. It is suitable for lines with small cycle times or different loading positions, or loading orientation at the stations, and is usually used for larger parts. A parallelisation of the processes is generally possible by changing the parts between stations using a robot. However, the processes are usually connected in series. The third system consists of a fixed linkage using a shuttle system. This is typically used between two stations and is characterised by transport using a workpiece carrier with multiple collets. This feeding system is suitable for medium cycle times between 5 and 30 s with medium part sizes and component weights of up to 5 kg. The accessibility is comparable to a typical line. The fourth system is a line, also referred to as transfer line, which is based on mobile workpiece carriers. In terms of feeding speed, it is not as fast as a rotary table, however, it enables feeding multiple parts through the stations and is therefore the standard in fully automated manufacturing lines. The choice and design of workpiece carriers is a key factor for the efficiency of the manufacturing machine.

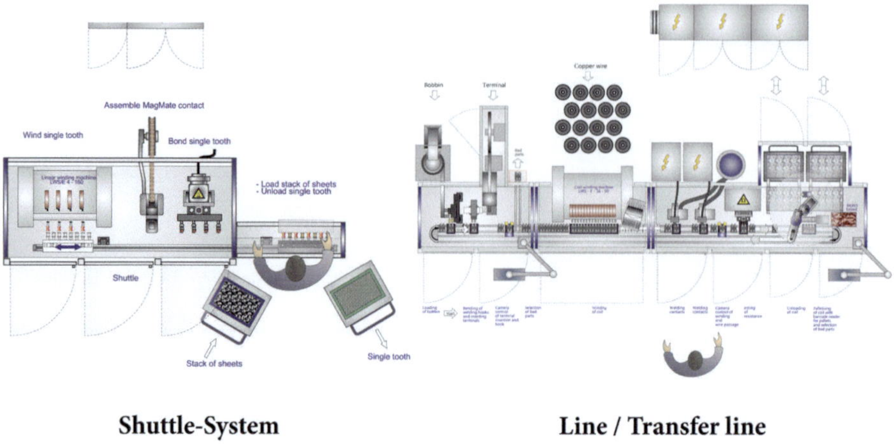

Figure 4.19 Layout concepts for feeding systems – linear systems

The system is defined by the product features (size, weight, surface), the desired processing situation (number and type of assembly steps), the required positioning accuracy for the assembly and the forces applied during assembly. The workpiece carrier defines with its collets the available positioning accuracy of the part for direct handling. Additionally, it is equipped with an encoding storage, often as a RFID (radio-frequency identification) transponder, in order to document the specific information of each product. In medical technology, for example, these have to be stored for decades. In terms of the structure, a distinction is made between conventional and flexible transfer lines.

The conventional transfer line is a rigidly chained manufacturing system for large series production, consisting of complementary machines. It produces a specific workpiece in a clocked production line. The manufacturing machines are arranged in series and simultaneously working on an individual process step. The cycle time is determined by the longest operation, which is usually the winding process. Due to the rigid structure, an optimisation of specific products is feasible, which in principle can achieve shorter cycle times. The main disadvantage of this layout is the behaviour during machine malfunction, because the whole transfer line comes to a halt. The overall availability results from the product of individual availabilities on the stations. A higher setup effort is required when changing to a different product due to the smaller flexibility. However, the advantage lies in special machine engineering. Each machine is specifically adapted to a task and can therefore work very efficiently. Transfer lines therefore have an extremely high productivity.

Due to their low flexibility and their high investment costs, typical rigid transfer lines are only economically feasible in the mass production of products with a lifespan of multiple years. As the number of variants increases, flexible multi-machine systems or flexible transfer lines are more and more replacing conventional transfer lines, and the product lifecycles become shorter.

On flexible transfer lines, different, but geometrically similar workpieces are manufactured ina cycle. The flexibility is achieved by using machines with NC-axes. Flexible transfer lines enable the economic manufacturing of part spectra, with production involving the same machine stations in the same order. They are suitable for processing of part categories in large series production due to the high flexibility and quick retooling. Typical processing examples are coils with the same geometries, but with a scalable length, depending on the desired power of the application. On flexible transfer lines, modular numerically controlled processing machines are used instead of single-purpose machines with predefined tools. As in conventional transfer lines, the material flow is cycled. A comparison of the different advantages and disadvantages of rigid and flexible transfer lines is illustrated in Table 4.4.

Table 4.4 Comparison of transfer lines

Concept	Pro	Con
Conventional transfer line	– Small space required – Simple logistics	– Reduced equipment availability
	– Small space required – Simple logistics – Constant manufacturing quality – Lean manufacturing (storage)	– Cycle time defined by slowest process – High setting efforts / low flexibility for different products
Flexible transfer line	– Higher flexibility regarding products	– Rare parallel manufacturing → Overall efficiency < 85 %
	– Higher equipment availability	– Output < 70 % based on cycle time analysis

Similar to conventional transfer lines, the individual processing machines are arranged in a row. They can be complemented by partial lines which serve for product specific processes, and buffers for elastic interlinking (Section 4.2.3). When designing a production machine, the cycle time analysis is a common tool used to gain an initial overview of the individual station properties. As previously mentioned, the winding process often takes longer than the downstream systems. By way of comparison, 180 s can be assumed for stator winding, where subsequent steps like contacting and testing can be performed within 4s. Due to this discrepancy, mechanical superstructures have been developed to separate the manufacturing, loading and unloading, such as pivoting frames for linear and needle winding machines.

In order to resolve the bottlenecks in cycle time and processing in winding technology, the number of spindles in winding machines can be increased, or multiple winding machines can be used in the line. This is partially also possible for other processes. A pick and place handling process can be accelerated with a stronger motor, while a welding process can hardly be optimised in terms of process times.

Aside from the concept of arranging the production machines and their chaining, the material flow to supply the machine with production resources, such as semi-finished products like contacting pins, or auxiliary materials like adhesives, also needs to be considered. The line concept provides good accessibility from the sides, so that the material supply for each station can be implemented from one direction.

If a customer requires flexibility in the production volume, a gradual upscaling can be planned during the layout phase. For the smallest stage, certain processes are planned as manual workplaces or processes are only partially chained, in order to implement a full interlinkage in the next stage. If a module is fully utilised during its cycle time and an upgrade is planned, bypass systems are often used to complement this module with further stations (Section 4.2.3).

It can be helpful for the planner to structure the configuration levels backwards. Beginning with the final configuration, a previous level can be created by removing feeders and stations, or replacing automation with manual workplaces.

During planning, the manual workplaces should be spatially aggregated if possible. This can be achieved by folding, crossing or mirroring of transfer lines, while taking into consideration the assembly order. An intelligent deviation from the linear layout or the integration of ejection channels can significantly shorten the distances between the individual manual workplaces, if the process chain contains multiple changes from manual and automated processes.

In order to put the costs of machine upscaling into perspective with its production volume, it can be said that another investment of 50 % of the initial machine costs can double the production capacity. Alternatively, the initial line can be re-built as a duplicate. This increases the overall machine availability, as both lines work separately. In this case, re-investment amounts to 80 to 90 % of the initial investment. An example for machine planning of a line with optional upscaling is illustrated in Figure 4.20.

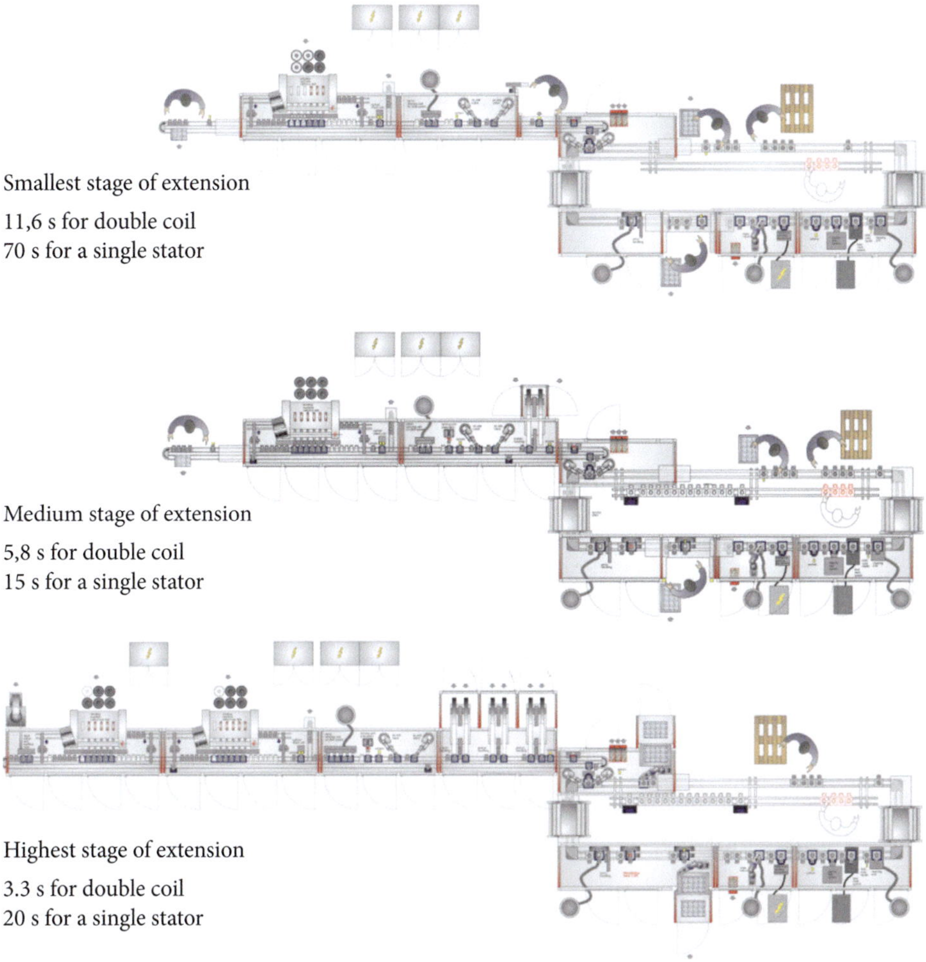

Figure 4.20 Possibilities for output increase / equipment extension

The objective of a manufacturing line is the highest possible machine effectiveness in terms of OEE (Section 4.1.3). There are different philosophies to achieve this objective, which ensure an effective organisation and an optimal operation. By way of example, lean production should be mentioned here. This describes a system of measures, which prevents waste in a production system in the shape of over production, stock in the system, long transportation or waiting times, ineffective production processes and rejects. For example, one specific measure is referred to as one-piece flow. This organisational approach is advantageous due to a high flexibility of workers, reduced delivery times and stock. The better qualification of the worker and the knowledge of the whole process chain results in a better product quality. This approach is known as the *Toyota Production System*. These approaches are applicable for every production system and not only for winding products. Therefore, this and other approaches can be found in standard literature.

The choice of a feeding system with sufficient repetitive accuracy for the assembly processes is important during the design phase. Transfer systems, for example, are limited to 0.03 mm, which is not sufficient for several applications in medical technology. Accordingly, the interlinkage may also be the limiting factor for the product quality or the achievable component requirements. A real manufacturing line layout for electric motor production is illustrated in Figure 4.21.

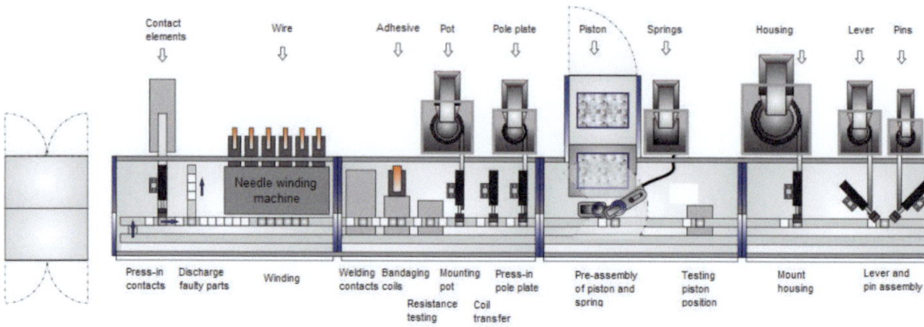

Figure 4.21 Example layout for a motor manufacturing line with needle winding

> The location is crucial for the preparation of a concept. For Western Europe, a machine with a high degree of automation is necessary due to the higher wages. Besides determining the degree of automation, the delivery conditions of semi-finished products as bulk material, on a pallet or as a blister belt is also important. The size of a product and the required tolerances must be considered for the handling technology. Rotary tables are preferably used for smaller parts, because they are more precise and the number of workpiece carriers is lower. The workpiece carrier tolerances are therefore known for a rotary table. The required production volume and the cycle times determine the type of automation of the interlinkage. Transfer systems are not used for cycle times below 1s.

4.2.3 Types of interlinkage for manufacturing lines

The technological possibilities of interlinking manufacturing lines have already been discussed in the previous section by introducing rotary tables, shuttles with workpiece carriers and transfer systems. Therefore, this section focuses on the organisational level with the material flow. In principle, a distinction is made between rigid, elastic and loose linking.

Rigid interlinking is characterised by the simultaneous transportation of all workpieces in the system. A new cycle only begins, once every assembly operation is finished. Accordingly, this system has the longest cycle time. An example of rigid interlinking is the rotary table. The advantage of this system is low stock in the production layout, so that it complies with the lean approach. The disadvantage is that the overall availability directly depends on the individual availabilities of the process elements involved. As opposed to this, in elastic interlinking the individual workplaces are decoupled, so that a failure of one station does not necessarily stop the whole system. An example for this is the transfer line. The advantage is a low failure susceptibility. The disadvantage is the need for buffers in the production system. Lastly, a loose interlinking describes the transportation of workpieces on demand. As a result, the elements are completely decoupled from the manufacturing process and transportation. A typical setup of this system is a transfer system with a bypass.

The most important properties of these three types of interlinking are summarised in Table 4.5. To choose a type of interlinking certain factors, such as stock, desired flexibility, availability of individual process elements and the cycle times, are crucial.

Table 4.5 Comparison of interlinkage types

Interlinkage Criterion	Rigid interlinkage	Elastic interlinkage	Loose interlinkage
Schematics	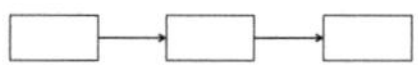	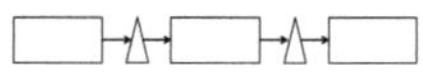	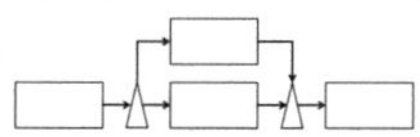
Cycle time per station	Equal	Nearly equal	Flexible
Storage	None	Buffer	Buffer
Time of transfer	Cycle time dependent	Cycle time dependent	Cycle time independent
Fault liability	High	Medium	Low
Sequence	Rigid	Rigid	Flexible
Processing time	Small	Medium	High
Stocks	Small	Medium	High

Interlinking with a bypass, using two machines with the same number of spindles, is chosen for multi-spindle systems providing a partial output, even if one machine fails. Due to the small cycle times for flyer winding technology, often a rotary table with rigid interlinking is used for process parallelisation.

The cycle times of the winding process dominate the manufacturing system, which is why a rigid interlinking is rarely chosen for winding machines. By choosing automated multi-spindle machines, higher production volumes can be achieved, although the cycle times are higher. This makes testing for winding readiness and ejecting reject parts easier.

4.2.4 Degrees of automation for winding machines

The degree of automation is defined as the ratio of operations with an automated production machine as part of the total number of operations and is mathematically described as follows:

$$Degree\ of\ Automation = \frac{Number\ of\ automated\ operations}{Overall\ amount\ of\ operations} \tag{4.11}$$

In terms of winding technology, as introduced in Section 4.1.5, three degrees of automation are distinguished. These will be described in detail in this section. Basic mechanical setups in the shape of universal winding machines, or table top winding machines have the lowest degree of automation. These are intended for small lot sizes and production volumes. The machines only consist of a winding spindle and a placing axis to position the wire guidance. Only the actual winding process is automated. The machine operator performs the steps of mounting the bobbin, terminating the wire, placing the start and end wire, starting the winding process and removing the finished coil. The machine operator also conducts the actual process monitoring (wire tear).

Semi-automated solutions are designed for the manufacturing of medium sized series. The wire placement is based on a Cartesian system of coordinates for three linear axes (wire guidance in all three spatial axes). The termination and the winding on the bobbin may take place automatically. The machine operator performs the steps of mounting the bobbin, starting the winding process and removing the finished coil. The fully automated solution, the automatic winding machine with automatic loading and unloading, is suitable for large lot sizes and high production volumes. Attaching upstream and downstream processes (injecting pins, soldering, testing, etc.) usually takes place with transfer lines. The operating personnel is responsible for troubleshooting and assuring the supply of wire and coils. A comparing categorisation of the different degrees of automation is illustrated in Figure 4.22.

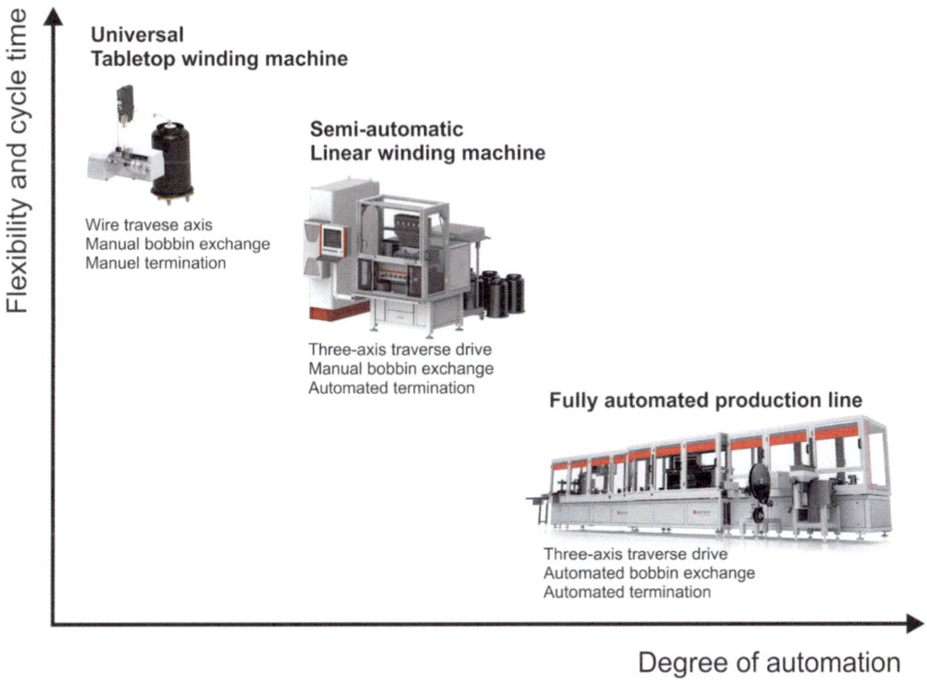

Figure 4.22 Different degrees of automation for winding machines according to [Dob-11]

> The degree of automation mainly depends on the production volume. However, during the planning phase of winding machines, the product life cycle should be considered, to avoid follow-up costs because of a high investment due to a rigid automation.

4.2.5 Part properties relevant to automation

Due to the large variety of product designs, only the most important features regarding automation practice can be covered here. As previously discussed in Section 3.3.11, the bobbin has a significant influence on the winding process. Accordingly, a lack of dimensional accuracy or stiffness may lead to errors in an automated winding process. Elements of wire positioning, such as the location of grooves on the bobbin ground or the wire inlet slot, are also important. A possible burr on the bobbin, as displayed in Figure 4.23, can cause abrasion to the wire and disturb the winding process.

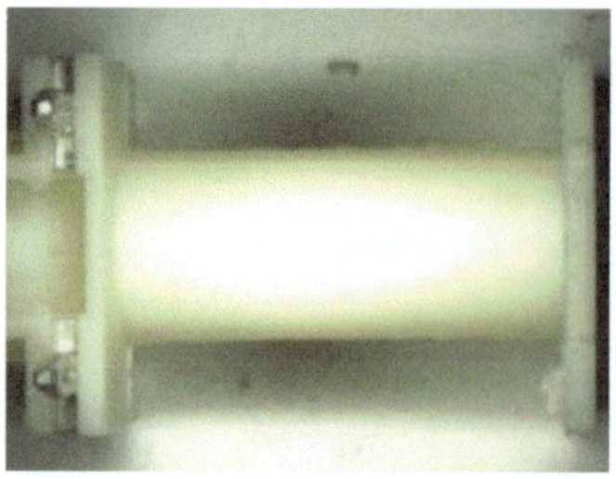

Figure 4.23 Typical error cause: burr on the injection molded bobbin

Generally, the manufacturability of bobbin geometries is challenging. They are usually manufactured by injection molding, which provides the best possible geometrical freedom. As an alternative, 3D printing bobbin is possible, but the process is not suitable for large series production. The limitations of the manufacturing process must be clarified before finalising the bobbin design. Depending on the bobbin design, the right choice of tolerances has to be considered. Firstly, the overall amount of the wire gauge tolerances in one layer may not exceed the winding width on the bobbin. Secondly, the summarised tolerances of layered wires may not exceed the flange height. Thirdly, the material strength of the flange must offer sufficient stability for the winding pressure. A product design suitable for automation should provide features for handling operations, for the feeding system, for decollating and should be stowable. Additionally, it has to provide features to detect the orientation as well as mechanical features for form-fitting torque transfer in the machine. If one of the previous aspects was disregarded, measures of error avoidance and correction can still be initiated. The best solution would be to improve the product design, in order to correct the original design flaws. As an alternative or in addition, further test steps could be implemented. This includes testing the parts for winding readiness before the winding process, monitoring the winding process and an extensive coil test with optical or electrical means of testing.

The Japanese philosophy *Poka Yoke* for precautionary error avoidance, which implements the intelligent design of part features, is also possible in winding technology and is already in use. This includes the use of defined flange heights from the inside and outside, grooves or marks on round parts for an error-free orientation. These marks are required for winding on the bobbin or terminating the start or end wire. Holes for pin assembly are very simple, yet effective features to avoid mistakes in the assembly of further elements. In the case of a higher product variety, different bobbin colours are used to differentiate product variants, based on the different wire gauges for example. Lastly, the position of the start and end wire can be named, for example in mounted coils. An example of a Poka Yoke feature for error-free orientation is also illustrated in Figure 4.24.

Figure 4.24 Poka-Yoke features for bobbins

A suitable bobbin design is crucial in terms of the process error susceptibility. The tolerances for semi-finished products may also have different influences on the winding result. A lack of bobbin stiffness may lead to its deformation. If the winding space is too narrow, the structure of the orthocyclic layering may be disturbed.

4.3 Automation techniques

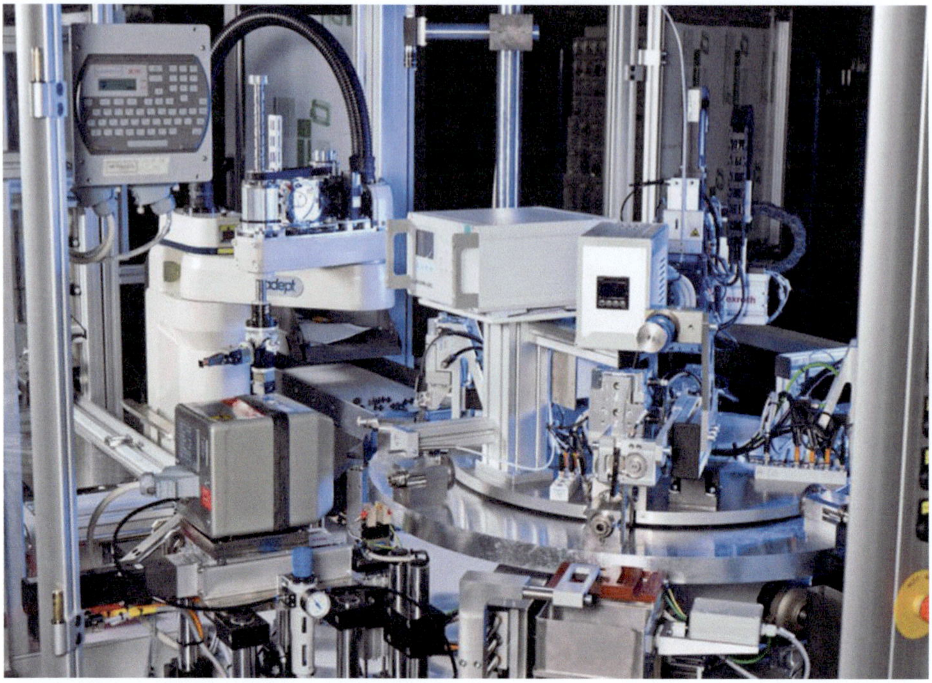

Figure 4.25 Station with rotary table and periphery

After focusing on the machine level in Section 4.2, this section covers the design of automated stations and their machine components as well as the design of manual workplaces.

Therefore, this section is directed at operators and developers for machine technology of production machines, as well as the manufacturing planning.

4.3.1 Common automation solutions

The purpose of this section is to introduce mechanical automation solutions and characteristic parameters, which are used for the manufacturing of winding products. The provided values are only guidelines. The following tasks are introduced as the central functions in winding technology:

- feeding
- assembly
- testing for winding readiness
- loading and unloading of winding machines

- contacting
- testing
- removal from the manufacturing line.

The feeding technology depends on the supply conditions of the semi-finished products. An oscillating conveyor with a feeding rate of up to 60 parts per minute can be used for bulk goods. For lower speeds, the use of an any feeder with a feeding rate of 30 parts per minute is possible. A palletiser, which can handle up to 60 parts in 5 seconds, can be used to feed pre-assembled parts. A bar magazine with a feeding rate of up to 60 parts per minute is an alternative to this. The structure of different feeding systems is illustrated in Figure 4.26.

Palletiser | **Adapt anyfeeder** | **Rotary feeder**

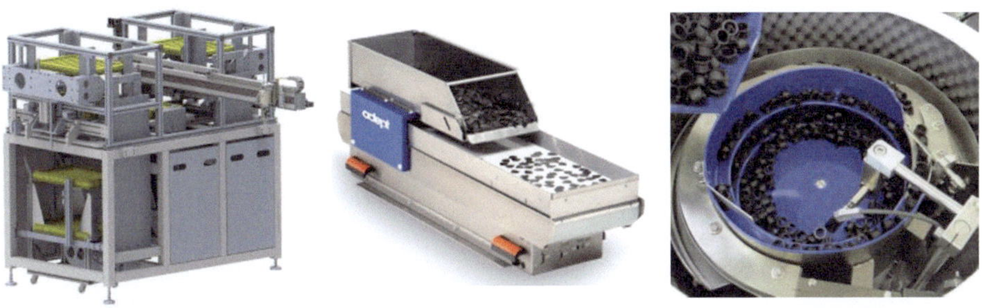

Palletising with a robot | **Handling from anyfeeder with robot** | **Palletising with linear axis**

Figure 4.26 Technical solutions for feeding systems. Scan QR code to view demonstration videos.

The assembly steps before winding can consist of an exemplary process chain of inserting pins into the bobbin and insulation papers into the slots. Inserting pins consists of a handling operation and a press-in process. The handling is typically implemented with a pneumatic gripper with a pivoting arm for cycle times below 2 s. Pressing in can be performed with a pneumatic press or a spindle press with servo or direct drive. Typical pressing forces are between 50 N and 500 N with an indent stroke of up to 10 mm. The repetitive accuracy of this process has a depth scattering of 0.1 mm at cycle times below 1 s. In order to insert the paper into the slot, this must first be folded and subsequently cut. After handling, it is assembled. The folding takes place using a set of stamping rolls with a length of up to 100 mm in less than 1s. For the separation, knives with a cutting width of up to 150 mm are used to cut the insulation paper in less than 200 ms. Alternatively, lasers can be used for folding and termination. These are significantly more flexible, although they are more expensive to purchase. The known pneumatic grippers with pivoting arm are used for the handling. The same applies to assembly, which can be performed with a repetitive accuracy of 0.1 mm in less than 1 s.

Tactile or visual testing systems are used for the test of winding readiness. The tactile testing takes place by means of a caliper with a measurement accuracy < 0.05 mm and a process time of under 1 s. Camera systems, contour sensors or light barriers can be used for the visual testing. Both have similar process times of less than 200 ms. The cameras have a better measurement accuracy of < 0.05 mm as compared to < 0.1 mm of other systems, however, they are much more expensive.

For the loading and unloading from the winding machine, pneumatic grippers, NC servo axes or robots (for very large bobbins) can be used. Both the pneumatics and the robot have a slightly worse repetitive accuracy of 0.1 mm as compared to 0.05 mm of the servo axes. The process times for NC axes and robots are slightly better with under 3 s as compared to < 4 s for pneumatic handling.

Three exemplary processes will be discussed for the contacting: resistance welding, insulation displacement contacts and soldering. Welding requires either pneumatic or servo driven welding tongs. Both operate with a welding current of up to 5 kA and accordingly have a sink-in depth of approximately 0.2 mm. However, the servo motors enable a quicker process in less than 1s and provide the possibility of a finer force regulation of up to 500 N. A contacting with the help of insulation displacement contacts requires handling for positioning, for instance using a pneumatic gripper, which typically takes up to 4 s and has a repetitive accuracy of 0.1 mm. The subsequent pressing can be performed with a pneumatic or spindle press. The required indentation depth can be up to 5 mm, which is achieved with a repetitive accuracy of 0.05 mm. The pressing force varies between 50 and 500 N at a process time of around 1 s. The soldering can be done by dip soldering or selective soldering, which also referred to as wave soldering. Dip soldering is achieved by dipping the contact into the soldering tin.

This takes up to 2 s, with the temperature of the tin bath being between 230 °C and 500 °C, depending on the wire insulation. For selective soldering, respectively wave soldering, the soldering tin is fed to the contacting location with a nozzle. Due to the connecting via soldering, the process parameters are similar to those of a soldering bath. However, the advantage is that the fumes, which are released from the bath, must not be sucked off separately.

The electrical testing can take place with different processes, although they can be performed on universal coil testers. For the testing, the conductor is contacted with a pneumatic gripper. Several exemplary properties of testing processes are summarised in Table 4.6.

Table 4.6 Automation relevant characteristics of testing processes

Testing process	Process time	Accuracy
Resistance measurement	300 ms	0,03 %
High voltage test	1 s	3,00 %
Voltage impulse test / Search test	500 ms	2,00 %
Partial discharge test with constant voltage	1 s	3,00 %
Partial discharge test with rising voltage	up to 1 min	1,00 %
Inductive response / Impulse response	1 s	0,05 %

Unloading the finished and tested coil is determined by the application background and can vary from palletising, unloading as bulk material into bar magazines, or accumulating conveyor. In this case the handling takes place using pneumatic modules or robots again. Because it is the most cost-effective solution, handling is often realised pneumatically. It is implemented, if the workpiece position is known, in order to enable the motion between two defined points. A robot on the other hand, provides a much higher flexibility in terms of picking and placing locations. A comparison is illustrated in Figure 4.27.

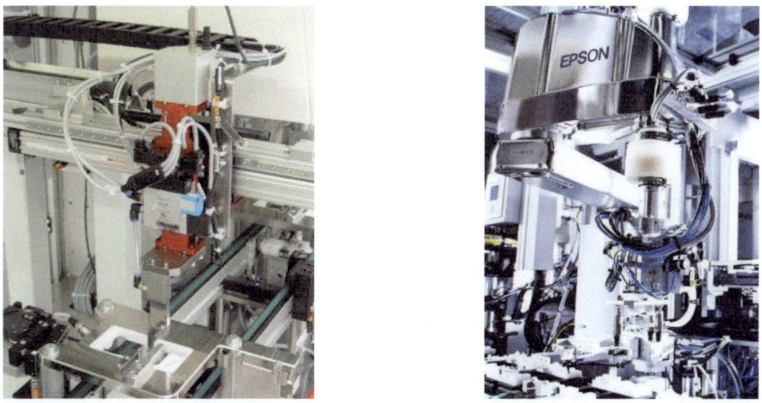

Figure 4.27 Comparison of handling with pneumatic gripper and robot

Typical automation solutions can be found in the handling, assembly and testing of parts at high production volumes. The machine technology is chosen with regard to the required cycle time. Handling tasks in different processes are usually performed using the same machine technology.

4.3.2 Integration of manual workplaces

An approach to make the production process more flexible is the consolidation of multiple tasks into a manual workplace. This might be due to changes in the order of the process steps or the performance of preparing operations, such as loading a workpiece carrier. For rising production volumes, a consolidation of different stations at one location makes sense, so that one person can operate multiple stations. Establishing this in the area of critical processes is useful because it enables the operator to react quickly. A possible integration of reworking workplaces to support the automated manufacturing or preparing operations is also possible, and suitable for series processes that still lack stability. There are different criteria to decide whether to integrate manual workplaces. Primarily, these are costs, so that the local wages and the number of shifts have to be considered. This also applies to the costs of machine technology, such as feeding technology, including the whole amortisation calculation. Additionally, the workplace has to be checked for possible hazards to the worker. This aspect is covered in more detail in the following section. The process quality is another criterion, as certain processes, such as the cable assembly, cannot be automated at all or only with great difficulty. A criterion with increasing importance is the aging workforce, which requires the implementation of ergonomics. The standards to be considered during the design are ISO 26800:2011 *Ergonomics – General approach, principles and concepts* and ISO 6385:2016 *Ergonomics principles in the design of work systems.* Figure 4.28 shows several examples for areas of stress during a standing activity.

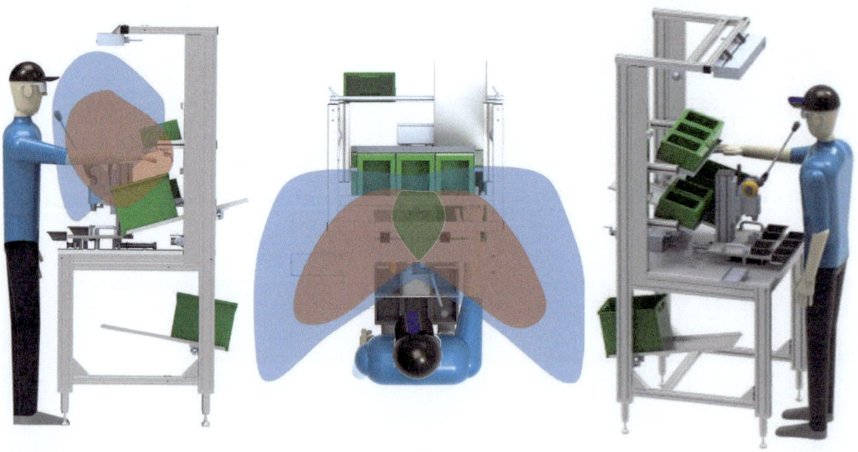

Figure 4.28 Ergonomic design of a manual work bench

Possible sources of hazards at a manual workplace or for a machine operator are high temperatures (risk of burning), fumes, liquids, electricity, machine motions and light. For example, high temperatures can result from contacting, baking of wires or hot folding of insulation paper. Suitable counter measures are heat extraction with suction, isolation or spatial separation, or a temperature monitoring with an alarm before the manual handling.

Health threatening fumes arise during contacting and can either be extracted, or automated by a feeding mechanism for the medium. Liquid hydrofluoric acid is employed in contacting processes, and splashes can occur in glueing or welding. Possible counter measures are isolation or cartridges used as a liquid container. An automated feeding mechanism for the adhesive, or the monitoring as well as an enabling mechanism that allows nozzle use only when it is at the station, can be used to design a safe bonding process.

For electricity, high voltages or high currents and the electromagnetic compatibility of machines have to be considered. Electrical protection mechanisms are the use of protective switches, a galvanic separation of machine components, as well as the insulation of live elements or grounding of operating elements and tools.

The dangers resulting from the motion of machine axes are already well known from machining manufacturing. In winding technology, they exist in winding axes, linear axes, presses or tools, such as cutting blades. Therefore, many protective measures are possible, such as light barriers, isolation by protective tunnels, stepping mats or protective panels, lower kinematics in the reach of humans, the installation of limit position switches, two-hand operation or machine locking.

The hazard from light that is associated with welding, for example with laser, must be protected against according to laser protection regulations. These include isolation, the installation of visual protection barriers according to laser classifications and the use of beam dumps with labyrinth technology.

Most common manual operations are used in the handling of bobbins for loading and unloading the winding machines, and secondarily for the preparative manual tasks like a pin assembly. For a concept, the use of an operation station depends on the cycle time. Generally, the limitation is a cycle time of 3 s. For a modular machine, different manual operations can be planned; these can then be replaced by automation in steps as the quantities increase.

4.4 Operation of automation

Figure 4.29 Stations connected via conveyor belt

Different boundary conditions and requirements must be fulfilled for a production operation of an automated machine. These will be introduced from a machine supplier perspective. They include the quality, process and machine capability.

The targeted readership of this section are machine operators for winding products.

4.4.1 Quality control

In this book, quality control for automated production systems is limited to typical manufacturing metrology and process monitoring for winding products, as well as the handling of this information in a quality management system. Key areas for quality parameters are electrical properties and the geometry. A subject-related discussion for winding was presented in Section 3.3.11. This can be summarised as follows: an accurate layer structure generally leads to accurate coil parameters. This is why controlling the winding pattern using a camera has often proven to be sufficient and has been established. This is achieved by comparing the pattern of the upper layer to a previously defined pattern of reference.

For the orthocyclic winding scheme, as for all winding structures, the errors in lower layers of the winding will always show in the last layer, as the error is likely to continue in every layer. Accordingly, a product reference is defined as the target contour or geometry and is complemented by a tolerance range. Using the comparison of the camera images, faulty windings can be identified. An example is illustrated in Figure 4.30. A principal concept for quality control is that the data are stored either in the central process control or in the workpiece carrier. The storage format contains a numerical value or a simple categorisation as *in order* (IO) or *not in order* (NIO) depending on the product and the customer requirements. The typical quality parameters for each process have been introduced in Chapter 3. Common testing parameters are resistances, wire tensions, press-in forces and strokes, solder bath temperatures as well as leak tightness or cleanliness. Depending on the task, multiple parameters can be recorded and monitored in one process, such as in welding: current, force, duration, distance and sink-in depth.

Depending on the complexity of the process chain and the requirements of each process, product dependent testing states are defined. For a large number of small assembly steps, quality control after vital steps is suitable, rather than a final test, in order to eliminate reject material. The higher testing efforts usually pay off due to the lower production depth for reject parts. The central data collection for operating data and the tracking of individual workpieces can be followed with barcodes or RFID. These are usually directly connected to the manufacturing execution system (MES) or the enterprise resource planning (ERP) system. The machine operator receives a warning when a diffusion of quality parameters from the manufacturing is observed; these can generally be recognised by a critical number of reject parts. The machine operator has to confirm the notification. Alternatively, another worker, the chief machine operator, could be responsible for finding the potential source of error.

Figure 4.30 Monitoring of the winding geometry

Thorough quality control helps to identify, and thus eliminate, the source of machine effectivity losses. Usually it is the producer, and not the machine supplier, who has a main interest in quality control, due to its high benefits. Therefore, the process and production machines are constantly optimised, often in consultation with the machine supplier. As feedback, the supplier often receives potential improvements and can therefore design future machines more efficiently, in terms of reject or productivity, or more robust, in terms of manufacturing deviations. This feedback also creates a service business, in which the machine supplier guarantees a certain reject rate and independently analyses and eliminates sources of non-achievement.

To achieve the quality, there is a tendency towards increasing requirements. Although the design of machines always focused on the output, the part quality is becoming increasingly more important. The efforts in information technology to implement different tasks is growing significantly in terms of process and manufacturing control. The necessary operating data acquisition has a high significance in the operation of production machines.

4.4.2 Error susceptibility – process capability

The error susceptibility of an automated machine is directly linked to the automated process. If a manufacturing process is unsuitable for a part, it will not work with a successful automation either. Therefore, it is important to refer to the value of process capability c_{pk}.

Table 4.7 Capability of chosen processes

Superior Process	Process or Machine	Process capability	Typical error cause
Wire forming	Wire drawing/rolling	2	Contamination/wear
Wire varnishing	Felt-/nozzle application	2	Contamination/wear
Feeding	Oscillating conveyor	1.33	Contamination/wear
	Anyfeeder	1.67	Outer light source interference
	Palletiser	2	Imprecise positioning in storage
Workpiece transport	Worker chain	1.33	Human influences/faults
	Transfer system/line	1.67	Contamination/wear
	Rotary table	2	Wear
Handling	Scara robot	1.67	Imprecise positioning
	Portal robot	1.67	Imprecise positioning
	NC Servo module	2	Wear
Bobbin assembly	Spring assembly	1.33	Tolerances of parts
	Pin assembly	1.67	Contamination, de-adjustment, deviation on positions

Table 4.8 Capability of winding processes

Superior Process	Process or Machine	Process capability	Typical error cause
Winding	Flyer winding	1.67	Wear tear, pulley failure, bearing failure
	Linear winding	2	Wire tear, maladjustment of wire guide
	Chain winding	1.67	Wire tear, maladjustment or improper positioning accuracy of wire guide
	Needle winding	1.67	

From automation practice, a list of c_{pk} values and typical error causes was prepared for process specific error sources and their incidence. The values provided in Tables 4.7, 4.8 and 4.9 should not be generalised but serve as basis for identifying possibly critical processes during design, and for generating counter measures for the specific cause or error.

Table 4.9 Capability of downstream processes

Superior Process	Process or Machine	Process capability	Typical error cause
Contacting	Solder bath	1.33	Dross generation on solder bath surface, copper content in the solder bath, heat dissipation into the pin area
	Arc welding	1.67	Contamination of electrodes, positioning of wire and contact pin
	Wire bonding	2	Contamination of bonding unit
Joining	Glueing	1.33	Contamination of application nozzle, wrong mass regulation control
	Pressing in	1.67	Stuck parts, burr formation, maladjustment of force sensor
	Riveting	2	Rivet contour wrong
Testing	Camera check	1.67	Outer light source interference
	Force-stroke measurement	1.67	Maladjustment sensor position
	Electrical testing	2	Contact contamination, position inaccuracy of the parts, varying contact resistances
Marking	Pad printing	1.33	Contamination printing plate, printing head
	Ink jet printing	1.67	Contamination ink jet
	Labelling	2	Change in temperature or glue properties

The actual c_{pk} value is the limit of what is currently technologically feasible. Accordingly, it decreases, if product specific features interfere with the process capability. Therefore, these tables should be regarded as assistance for process design during the planning phase. The most common source of error in winding is the winding tool, or the product specific machine components. A reason may be the missing consideration of production aspects for product properties due to a lack of involvement in the product development. The size of the machine elements can lead to problems due to the miniaturisation of the parts, for example in bobbin handling, assembly of peripherals (pins) or generally in all processing steps, which include assembling/joining or separating.

Possible sources of error outside of the processes, meaning on the machine side, can easily be ruled out by a sufficient calibration of important components, such as the wire tension sensors. Maintenance according to the recommended servicing plan is also a prerequisite for functioning machine components. Figure 4.31 displays typical measures of calibration.

Wire brake calibration using a weight

Adjustment jig for linear winding machines

Figure 4.31 Calibration of machine components

When complying with the prescribed usage of machine technology, the error susceptibility lies in the process capability of the individual processes. Exceptions to this are process disturbances due to sporadically occurring deviations in the production environment. These are covered in the following section.

4.4.3 Error diagnosis

A process related diagnosis of errors can take place according to the procedure based on a fault tree analysis, or a preceding FMEA as described in Section 3.3.11. Accordingly, errors can be classified using their incidence. If they are constantly present, they are process related and can be assessed using the risk evaluation. Sporadically occurring errors, however, are much harder to characterise and to eliminate. Typical sources of error for sporadic errors are contamination or the wear of components. In wire guidance, wire abrasion can lead to an accumulation of lubricant. This applies for all wire guiding components. Other sources of error are often the sporadically varying deviations of semi-finished and part properties, which lie outside the tolerances. Changing properties of semi-finished products can be related to the delivery conditions. It is also possible that local environmental influences interfere with parts delivered in good condition and lead to process errors. For example, humidity: Storing certain synthetic materials at high humidity can lead to water absorption. In turn, this leads to swelling of the bobbins and simultaneous lowering of the stiffness.

Failure Mode and Effects Analysis — Machine / Process - FMEA																		
Product name:						Customer:						Version:						
Project No.:												Date:						
No.	Process description	Function	Failure type	Local effects of failure	(S) Severity	Potential cause(s) / mechanism	(P) Probability	Failure detection	(D) Detection	RPZ	Counter measures	Responsible person + deadline	Measures carried out	(S) Severity	(P) Probability	(D) Detection	RPZ	
1	Four workpiece carriers with stators are being cycled into the winding machine with a separation length of 180 mm	Bring workpiece carriers with stators into transfer position	Workpiece carriers do not enter	Stop cycle	1	Conveyor defect	5	Standstill of machine	1	5	Frequent maintenance of wear parts						0	
2			Workpiece carriers are not positioned accordingly	Parts are mounted on an angle	5	Defective workpiece carrier or centering	5	Downholder signals wrong position	2	50							0	
3				Stator unusable	6	Defective workpiece carrier or centering	5	Downholder signals wrong position	2	60							0	
4				Possible wire guide fracture	6	Defective workpiece carrier or centering	5	Downholder signals wrong position	2	60							0	
5			Already wound stators on workpiece carrier	Possible wire guide fracture	6	Operator places wound part on carrier	6	Machine stops after wire tear	1	36							0	
6				Stator unusable	6	Operator places wound part on carrier	6	Machine stops after wire tear	1	36							0	
7			No stators on workpiece carriers	Wire loss in needle carrier	2	Operator places no part on carrier	6	Machine stops after wire tear	1	12							0	

Figure 4.32 FMEA for error analysis

In error diagnosis, mechanical sources of error should be analysed first. Even if there are deviations on the bobbins, in most cases these can be traced back to their automated manufacturing. When locating the source of error, 90 % of errors have a systematic origin, regardless of whether this was caused by deviations in the winding spindle or by holes in the injection molded bobbins. Accordingly, all process steps have to be analysed and documented systematically before an error occurs.

4.4.4 Maintenance

The documentation for each production machine includes a maintenance plan for the machine elements, including a list of corresponding wear parts. Maintenance minimises the signs of wear on the machine components by providing regular care, thus increasing the lifespan of these components and the whole machine. The excerpt of a maintenance plan is shown in Figure 4.33. It contains information regarding the component lifespan, incidence of the activity, information regarding design features of the machine and references to the technical drawings. Failure to carry out these maintenance activities can lead to increased reject rates or even machine failure in extreme cases. Accordingly, in terms of the OEE, the machines should be maintained. It has been found that reactive servicing strategies, meaning acting when the machine has already malfunctioned, lead to lower machine output and provide no predictability for maintenance measures. The opposite, i.e., performing precautionary measures, is referred to as preventative servicing and enables optimal machine availability due to low failure rates. However, the machines must be available for servicing at regular intervals. Accordingly, this results in a higher workload and a higher demand for spare parts due to the premature replacement of components. The optimal strategy depends on a number of factors and should be chosen or changed as required.

Pos	Maintenance intervals	Maintenance tasks	Figure
3.) Basic configuration swivelling frame			**60.516.758.000**
3.1	See manufacturer's information	Flexo coupling FK M8 of the side pieces. *Check the coupling for function and wear and replace it if necessary.* see KTR manufacturer's information ***60.500.982.000***	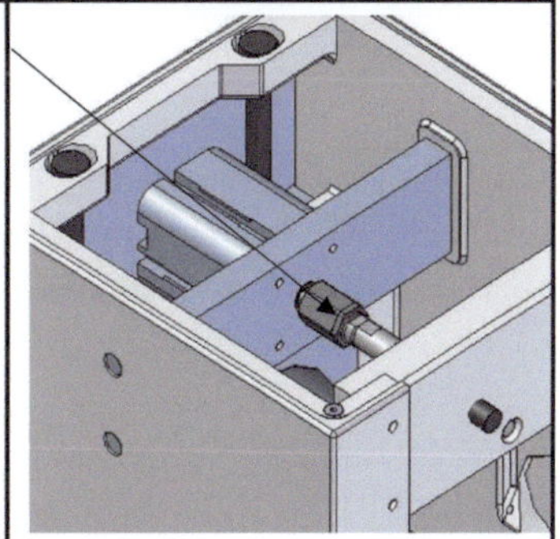
3.2	approx. 2000 operating hours	Toothed belt 5M-25-575 of the step gear. *Check the toothed belt for tension and damage, replace it if necessary.* ***AUMANN** recommends that you replace all belts of at least 5 years or 42,000 hours of operation.* ***60.500.983.000***	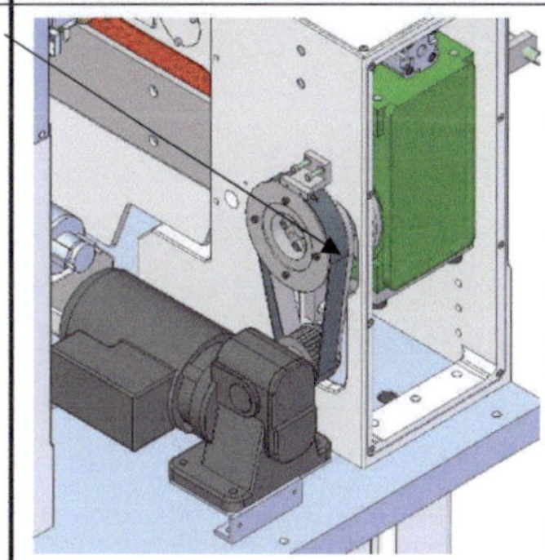

Figure 4.33 Extract from a maintenance plan for a swivelling frame

Components with the lowest possible maintenance efforts can be selected, in order to save maintenance costs in terms of the total cost of ownership (TCO) approach. In addition, the capabilities of the staff and their technical expertise for maintenance of different technical systems have to be considered when purchasing.

Maintenance of assembly lines for winding products is an important prerequisite for high machine availability. When compared to machining, these efforts are within reasonable limits.

4.4.5 Requirements for operating

The requirements for operating a winding machine are precisely defined in the documentation, and during handover. Each machine in a plant has its own documentation which contains

- safety regulations and instructions
- intended use and operating principle
- transportation and technical documentation
- electrical installation
- operating elements
- instructions on how to use the machine
- possible errors and their elimination
- information on maintenance and care
- part list incl. information about external suppliers.

The handover of the machine also includes training of staff. The machine operator is responsible for ensuring compliance with technical requirements (interfaces, maintenance, etc.) and specifications for environmental conditions (temperature, humidity). Only then, are the legally agreed terms for warranty period or production time incl. output binding. After the first training on the machine, it is also the operator's responsibility to have a sufficient and qualified pool of personnel for possible capacity bottlenecks. According to their expertise and tasks, the personnel can be categorised into the following groups: machine operator, system operator, service technician and material logistician. Each group has its own access rights and degrees of freedom in terms of the machine control. The machine operator is only responsible for the production procedure and cannot change any settings besides loading and setting up. The system operator can change the process and its control in detail, in order to be able to set up new production processes. The service technician sent from the machine supplier has full access, even for internal machine values, which are of no use to the machine operator. Lastly, the material logistician has the rights to supply material to the production material. The rights allocation is summarised in Figure 4.34.

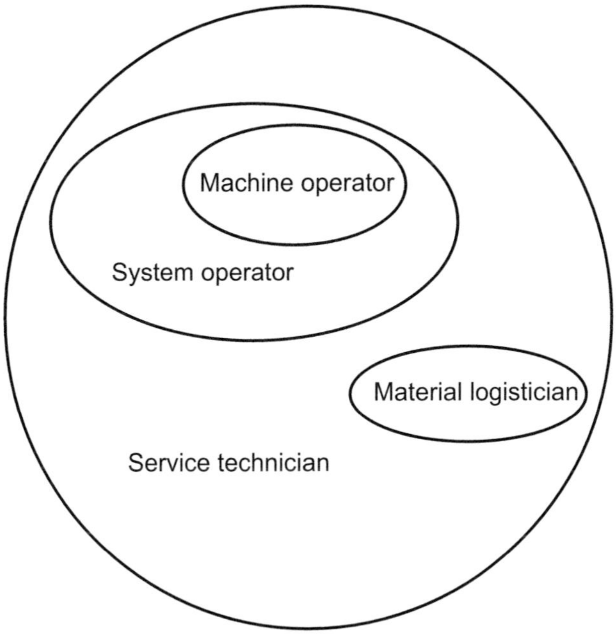

Figure 4.34 Access management for machine control

It is essential for the functional capability of the process chain that incoming production material is quality controlled, based on random samples, is essential and usually standardised. It is the machine operator who defines the characteristics of the quality control.

Besides maintenance, adjusting and calibrating of machine components is essential during the use of the machine. The use of setup gages and training how to use them enables the machine operator to swiftly set up the machine.

References

[Bo-04] Robert Bosch GbmH: Maschinen- und Prozessfähigkeit. Quality assurance in the Bosch Group, Technical statistics, 3rd edition 1.7.2004

[Br-11] Brunner FJ, Wagner: Qualitätsmanagement. Leitfaden für Studium und Praxis. Hanser 2011

[HES-93] Hesse S: Montagemaschinen. Vogel Buchverlag, Würzburg 1993

[HES-00] Hesse S: Fertigungsautomatisierung, Automatisierungsmittel, Gestaltung und Funktion. Vieweg 2000

[La-10] Langmann R: Taschenbuch der Automatisierung. 2nd edition. Hanser, Munich 2010

[Ro-14] Roth M et al: Standardisierungskonzept für Kleinserien im Maschinen und Anlagenbau. Conference contribution, Tag des Systems Engineering 2014. Bremen, 12–14.11.2014

Index

G

H

I

J

L

M

N

O

P

R

S

T

V

W

MIX
Papier aus verantwortungsvollen Quellen
Paper from responsible sources
FSC® C105338

If you have any concerns about our products,
you can contact us on
ProductSafety@springernature.com

In case Publisher is established outside the EU,
the EU authorized representative is:
Springer Nature Customer Service Center GmbH
Europaplatz 3, 69115 Heidelberg, Germany

Printed by Libri Plureos GmbH
in Hamburg, Germany